# Understanding Digital TV

**IEEE PRESS Understanding Science & Technology Series**

The IEEE PRESS Understanding Series treats important topics in science and technology in a simple and easy-to-understand manner. Designed expressly for the nonspecialist engineer, scientist, or technician as well as the technologically curious— each volume stresses practical information over mathematical theorems and complicated derivations.

Other published and forthcoming books in the series include:

**Understanding the Nervous System**
*An Engineering Perspective*
by Sid Deutsch, Visiting Professor, University of South Florida, Tampa and Alice Deutsch, President, Bioscreen, Inc., New York

1993   Softcover   408 pp   IEEE Order No. PP0291-5
ISBN 0-87942-296-3

**Understanding Lasers**
*An Entry-Level Guide, Second Edition*
by Jeff Hecht, Sr., Contributing Editor, *Lasers and Optronics Magazine*

1994   Softcover   448 pp   IEEE Order No. PP0354-1
ISBN 0-7803-1005-5

**Understanding Telecommunications and Lightwave Systems**
*An Entry-Level Guide*
by John G. Nellist, Consultant, Sarita Enterprises Ltd.

1992   Softcover   200 pp   IEEE Order No. PP0314-5
ISBN 0-7803-0418-7

**Understanding Electro-Mechanical Devices**
by Larry Kamm, formerly President of MOBOT Corp.

1995   Softcover   384 pp (approx.)   IEEE Order No. PP03806
ISBN 0-7803-1031-4

Ideas for future topics and authorship inquiries are welcome. Please write to IEEE PRESS: The Understanding Series.

# UNDERSTANDING DIGITAL TV
# The Route to HDTV

Brian Evans

The Institute of Electrical and Electronics Engineers, Inc., New York

This book may be purchased at a discount from the publisher when ordered in bulk quantities. For more information contact:

IEEE PRESS Marketing
Attn: Special Sales
P.O. Box 1331
445 Hoes Lane
Piscataway, NJ 08855-1331
Fax (908) 981-8062

© 1995 by the Institute of Electrical and Electronics Engineers, Inc.
345 East 47th Street, New York, NY 10017-2394

Printed in the United States of America
10 9 8 7 6 5 4 3 2 1

**ISBN 0-7803-1082-9**

**IEEE Order Number: PP04366**

Library of Congress Cataloging-in-Publication Data

Evans, Brian.
    Understanding digital TV : the route to HDTV / Brian Evans.
        p. cm.—(IEEE Press understanding science & technology series)
    Includes index.
    ISBN 0-7803-1082-9
    1. Digital television. 2. High definition television. I. Title.
II. Series.
TK6678.E93 1995
621.388—dc20
                                        94-20667
                                        CIP

# Contents

# Preface

For as long as I can remember the way that television works has held a fascination for me. My first experience of the television industry was gained in my school days when my summer vacations were fully spent in the local TV shop helping to convert everyones's existing television set to something they all wanted desperately—ITV.

The late 1950s was boom time for the TV industry. Independent Television, ITV, had just started its UK broadcasts and our choice of stations had doubled overnight. We were now able to enjoy two TV programs instead of just one.

My older brother had mixed feelings about this technical advance. On the one hand we were now able to watch both BBC and ITV on a modern 14-inch screen but, on the other hand, the bedroom which we both shared had become filled with television sets, all in various stages of disrepair.

Just occasionally he would summon up courage, look over my shoulder and ask what I was doing. I soon discovered that, however much I tried to explain the technical intricacies of what went on inside the television set, he remained remarkably unimpressed with my hard work. Unimpressed, that is, until one day I was able to repair his tape recorder.

We both learned from that experience.

He learned that there was some use after all in my dedication to things technical. I learned that, unless the technology could be easily explained and shown to be useful, for most of mankind it need not exist at all.

Thirty years later the UK legislators had to grapple with and try to understand a much wider range of technical issues. Broadcasting had changed a lot since Elvis Presley and Buddy Holly first appeared on the screen and now it came under the political microscope once more.

Two years before the 1990 Broadcasting Bill finally slipped onto the statute book, the UK Parliament asked their Home Affairs Select Committee to investigate the current state of UK broadcasting. This committee had the power to call witnesses and, for more than three months in early 1988, they heard evidence, highly technical evidence most of the time, from a whole line of experts. They listened in disbelief to one of their number, an ex-broadcaster, who asked the experts when they thought digital television (and high definition digital television in particular) would arrive in the marketplace.

His questions often elicited blank stares—what on earth was digital television?

In August of 1987 the American Federal Communications Commission had set out on a similar task. They had issued a Notice of Inquiry (an NOI) into the subject of Advanced Television for the USA and many research groups had responded to the challenge. But, back in 1987 the subject of digital television attracted the same blank stares.

It was not until 1990 that things began to change.

During the discussion phase of the UK Broadcasting Bill many MPs became worried. Digital television was now more than just a speck on the legislative horizon for it was mentioned regularly in the proposed amendments to the Bill. Many MPs who heard of these American "digital" developments wished out loud for up to date technical briefing rather than needing to rely on Select Committee evidence which was now two years out of date.

They feared that, without up-to-date information, any legislation put in place would be needlessly reactive rather than pro-active.

The origins of this book lie in the various briefing notes which I provided for parliamentarians of all political persuasions during the passage of the Broadcasting Bill through the two Houses of the UK Parliament.

The book is therefore aimed at anyone who would appreciate a generally nontechnical guide to what it all means. It strives to cover that information gap, the same information gap that, many years ago, kept my brother and me in different states of perception. Unless I can make him understand enough about this subject to realize that digital television is one of the most exciting ideas of the decade then he is going to miss out on quite an adventure.

It is written in the style of a detective story, complete with clues and false trails, plots and counterplots, heroes and villains. In reality I believe the following pages will prove far more enjoyable than detective fiction because the research I describe has been done by real-life detectives and we are free to determine the conclusions for ourselves. Unlike the detective story we are offered a plot in which there is a good chance that many competing interests will be able to prosper together.

The first chapter, entitled "Digits Galore," explains the "digital" part of television in nontechnical language while almost disregarding the high definition overtones until the very end.

The second chapter, entitled "Perceptions," sets us a major question. Is high definition television merely to be defined as a picture that is twice as sharp as today's TV pictures or is there more to it than that? The answer is not only fascinating but is even quite scary and it is easy to envision the commercial opportunities that can flow from these realizations.

The third chapter, entitled "Simply Irresistible," draws the threads of the first two chapters together. It shows how, with the aid of digital television technology, we can still construct a single worldwide HDTV distribution standard that can satisfy everyone—not in total but probably more than sufficiently to provide a win-win scenario that will ensure worldwide conformity. The last part of the chapter sets out seven practical scenarios that might offer a profitable path from today's conventional television systems to something much better.

This book is not only about digital television, it is also about many other aspects of digital broadcasting of which digital HDTV forms only a small but important part. It does not endorse any particular proprietary form of digital technology, preferring instead to outline the many different families of digital television and audio techniques that can be pressed into use. This book sets out the opportunities and benefits that digital television will bestow on all of us.

To take full advantage of these developments we will need to establish a strategic framework, a political and commercial understanding that will continue to bring these opportunities to fruition over the next 10 to 15 years. Without such an understanding we can easily fall prey to the street trader with his wheelbarrow full of brightly colored baubles and beads.

*Brian Evans*

# 1

# Digits Galore

**INTRODUCTION**

Not long ago I decided that I should improve my domestic Hi-Fi with the addition of a good compact disc (CD) player. I bought and read some Hi-Fi magazines and then visited some of the local Hi-Fi shops which had placed advertisements in the magazines.

I was soon confused.

Some portable CD players could be bought for about $150, whereas the housebound CD units cost anything from $200 to $1,500. What might be the difference in quality, and, perhaps more to the point, would my own ears and brain notice any difference between the different models?

I plucked up courage and visited a local dealer for a demonstration. I was rather taken off my guard by the salesman who told me that many Hi-Fi enthusiasts still preferred the musical quality of 12-inch long playing albums—he called them "vinyls"—to the quality of compact discs. I asked what basis there was for such an assertion but he was quite unable to explain why this happened.

We moved on.

I asked, "What is the real difference between an inexpensive $200 and an upmarket $1,500 CD player? Both CD players would appear to have nearly identical specifications which are printed in nearly identical glossy brochures."

"I can't tell you that," he replied, "because I don't know. You must listen and find the difference for yourself."

I winced.

Would I recognize the more expensive player as having better sound quality or would I be unceremoniously kicked out of the showroom once it was discovered that my ears were clearly not up to the task in hand?

I was seated in front of the equipment and was able to listen to the same piece of music, first through the inexpensive CD player then through the expensive one.

I swapped the CD back and forth a few times. After listening for a while I volunteered that there did appear to be a slight difference but I was still unsure whether it was worth my while paying the extra money.

At last the salesman took pity on me and let me into the secret.

"The difference," he explained, "is in the background sounds. From the cheap CD player you will hear the music clearly but from the expensive CD player you might also notice—after a little while—the musicians shuffling their feet."

I listened hard and, YES, the musicians did dutifully shuffle their feet in the expensive CD player.

So this must be what you pay extra for!

The inexpensive player appeared to provide music which was little different to the expensive player except that, just like the aftertaste you would expect of a good wine, the expensive player provides that subtle depth of feeling perhaps best described as "ambience."

I was delighted by the unexpected honesty of the salesman who had merely told me what his own ears had told him—especially as he did not know how it had been technically achieved.

The following day I telephoned the makers of the expensive CD player, complimented them on the unexpected audibility of the musicians' feet and asked them how they did it. The answer was surprising.

"Very accurate timing."

"But," I replied, "in the sort of digital circuits which are always used in CD players is not that sort of accuracy always assured?"

"Only if you are very, very careful with the layout and the design. What might at first be considered to be imperceptibly slight variations in timing can easily destroy the ambience behind the sound."

> Here at last was a technical insight into a familiar phenomenon—that concentrated listening allows us to hear things that were not evident, if we can mix metaphors, "at first sight."

"How large are the timing variations?" I asked.

"Less than a millionth of a second," came the reply.

I was shocked. Had I not always been told that it is our eye, not our ear, which is the marvel of nature and had I not learned that we see pictures on our television sets by virtue of signals that often last less than a millionth of a second? In contrast, I had been led to believe that our ear was perhaps the poor cousin to our eye, quite unable to even hear the allegedly deafening sounds made by ultrasonic dog whistles, garage door remote control units and bats.

How could the ear possibly hear timing variations of a millionth of a second?

I started to think afresh about our unexpected sensitivity to time and how apparently "imperceptible" timing variations can send confusing signals to our brain.

I asked myself whether our eye was indeed more sensitive to environmental variations than our ear or was it more likely that our eye-brain and ear-brain were running biological software programs of roughly equal power.

I found it useful to skip between two parallel trains of thought. The clues to good audio reproduction often led me to a better understanding of what might constitute good visual repro-

duction and hence to a better understanding of what high definition television might turn out to be.

But there was a catch.

A full understanding of the visual requirements of high
definition television presupposed that I had somehow acquired a
"critical eye" to analyze the picture. This was familiar ground.
Had not the CD player salesman kindly presupposed that I had
a "critical ear" through which I might hear the musicians shuffling their feet?

Not quite.

I must admit that I did not hear the background sounds until
the salesman told me to concentrate hard and to listen very
carefully for them. Then things changed. Once I knew that the
shuffles were already there, just waiting to be discovered, I was
able to "learn" fast and could soon hear them clearly. Clearly, that
is, on the expensive CD player but not at all on the inexpensive
CD player.

With all the will in the world I could not have developed a
critical ear by listening to the inferior CD player.

What was missing?

More important, what were the positive factors that had
allowed me to make good progress in training my ear when I
listened to the expensive CD player?

Was it something in the music itself that held the clue?

## Masks and whiskers

Classical music which is played on acoustic instruments
provides a good example of discovering hidden depths in both the
music itself and in individual performances.

Take Bach for instance. The rigid, strictly metronomic beat
of a Bach fugue allows us to build up in our mind's eye (the wrong
metaphor of course) a deeper appreciation of what, from the
published musical score, is an astonishingly economic use of
notes.

And yet, the full beauty of the Bach fugue cannot become
apparent unless it is played by a concert pianist or by a student
who has virtually mastered the piece.

It is only at that point of stability, where command of the

keyboard is now so complete that it can almost be discounted, that the music can and does suddenly take on a life of its own. Not only can we see much deeper into the music but we are also invited to gain insights into the mind of the composer and the feelings of the artist. This must be an example of human communication at its very best.

There are a number of factors that influence our appreciation. The first is the skill of the musicians in providing a stable interpretation. The second is the need for a stable environment in which we can listen—free of the noise of dogs barking and aircraft flying overhead. It is only when both these criteria are satisfied that the third factor—the "trained ear"—can connect us to the deeper nuances and subtleties that lie behind the music.

Music is not alone in its need for an educated audience. Many visual images can remain totally unappreciated unless we have the equivalent "trained eye" to detect what, at first sight, may appear to be unimportant and almost imperceptible changes in color, texture or contour.

For example, a change of a fraction of an inch in the cut of a dress can change it from chic to cheap—and, in so doing, easily effect a substantial reduction in the price of the garment.

How can these value judgments be made so swiftly and accurately? It is almost as though the experienced fashion conscious buyer carries with her a secret mental plumb rule which, in her mind's eye, can be set against the dress in order to assess its worth.

We experience this plumb rule effect more directly when we try to hang a picture on a wall. If the wall is covered in a decorative pattern—especially if it contains horizontal or vertical repetitions—then it takes great patience on our part to set the picture level on the wall.

The reason for our difficulty is plain—the wallpaper has provided us with a ready-made two dimensional grid or mesh against which it is difficult to stop ourselves forever checking and correcting the alignment of the picture.

In the language of the computer industry this wallpaper reference grid or mesh has been implemented in "hardware" whereas the fashion buyer's application of her mental plumb rule is an alternative more flexible "software" implementation.

We can develop this idea of a software mesh a little further.

Perhaps the phrases "trained ear," "trained eye" and "mind's eye" are no more than expressions of our inner ability to construct a software reference grid in our mind against which we can then exercise critical judgment. But is it not curious that the word "trained" has another, subsidiary, meaning? We know that we can "train" our eye to ever greater heights of critical judgment yet we can also say that we "train" our eye on a distant target. Both meanings of the word imply discipline and steadiness or stability in their implementation.

Stability is very important.

A pair of high-power binoculars are difficult to use in a moving vehicle and will only provide really good pictures when safely mounted on a tripod. Without the tripod the binoculars are next to useless—the slightest vibration results in a disproportionate loss of visual acuity. Home movies which are taken from hand-held cameras suffer the same apparent loss of sharpness, even though individual frames of the picture might have remained in sharp focus.

Why is this?

Let us return to the music for a moment.

We know from experience that, in order to enjoy the Bach fugue to the fullest, our own mental reference grid or mesh will take a few seconds or more to establish itself in our mind. Is it merely a coincidence that this mental reference grid software takes the same few seconds to load as does our favorite software application in a personal computer?

We can press the software analogy a little further. Writing the original software application code probably took the software company many man-years of effort and cost millions of dollars. In the same way it had taken us all a lifetime to build our own personal mental reference grid from scratch.

Can we recall how we did it?

Our individual reference grid started life as a fragile and flimsy mesh through which we first learned to both listen to and look out onto the world. It was at this stage that a temporary lapse in our childhood concentration could easily undo all our efforts in building and establishing the grid.

The passing of time has not made the learning task any

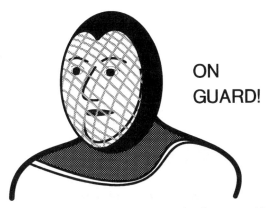

ON GUARD!

**Figure 1-1** Our personal mental reference grid.

easier. We know that extending the mesh takes a lot of concentration which can only be fruitfully undertaken in a stable environment. At times the stability demanded of the environment can verge on the unreasonable. I know that settling down—another interesting metaphor—to write a letter can temporarily place quite unreasonable demands for peace and quiet on my wife, my children and their friends.

Once set in place, however, the mesh gradually becomes stronger and more robust. It no longer falls such easy prey to those seemingly inevitable lapses in concentration but allows us to switch more easily into a much deeper perception and understanding of what we see and hear. Our individual reference grid is now wrapped around us just as a swordsman would don a protective mask. It helps us see and hear but also heavily influences how we perceive and interpret these signals. More often than not it determines what we come to believe.

Without this reference mesh in place we would never get beyond first base. We would be as lost as a cat without its whiskers. We would remain unenlightened and uneducated. In short, we would always be stuck with first impressions and effectively be left "in the dark."

However, our reference mesh differs from the swordsman's mask in one important respect.

Whereas the swordsman's mask serves to exclude and protect, our mental "cognitive" reference grid also enables us to draw on our past experience and thus to enhance our understanding of what we see and hear.

This process of education and enlightenment in the musical and visual arts requires three elements:

1. A society that nurtures and encourages excellence, thus providing us with a reference point—a bedrock of virtuoso musical performance or "old master" pictures to which we can all aspire.
2. A stable transmission medium which faithfully extends these images and performances from the studio or concert hall to everyone's home.
3. An audience of quiet minds who have welcomed the opportunity to appreciate and learn from such excellence.

But surely this is all self-evident and already in place? Art galleries and concert orchestras are to be found in every large city across Europe and the United States. There is a ready audience for good orchestral performances and exhibitions of fine art. Classical music can be heard on the FM wavebands. Agreed, these radio stations might not make an outright fortune but they seem to get by.

Where is the problem?

The problem lies in how we tackle the middle item—the link from the studio to the home. There is a danger that, if misapplied, digital sound and digital television techniques might become a long-term disadvantage rather than the blessing we all hope for.

The danger is not in the same league as the "Death Star," that delightfully emotive label which describes a proposed digital television satellite which is intended to beam 500 television channels to homes across America. The misapplication of digital technology is already with us in the shape of the $200 CD players which muffle the sound of the musicians shuffling their feet.

Why is the transmission medium so important?

We know that until 70 years ago the audience had no choice but to travel to the performance—the middle element in the above trilogy did not then exist. First radio and then television brought sound and moving pictures to our homes. Everyone's horizon was expanded. The quality of sound reproduction improved by leaps and bounds, eventually leading to the creation of that much abused marketing slogan—High Fidelity Sound.

We are now on the brink of another revolution in broadcasting. Yet today's impending change from analog to digital broadcast transmission is greeted with both enthusiasm and caution—some even fear its introduction.

Why is this?

Despite the inherent advantages of digital transmission, which are explained later in this chapter, there is great concern that, in our rush to "digitize" everything, we might inadvertently lose that elusive "essence" of image and sound.

There is a danger that we might easily throw out the baby with the bathwater. At first sight the transmitted audio-visual signals might still appear "squeaky clean" to the broadcaster but are we, inadvertently, discarding that secret ingredient which, in former times, would have turned us on?

Like most things in life—but for engineering in particular— what we get for our money is almost entirely dependent on our original specification. If we get the specification wrong then the results will be unsatisfactory.

The flipside of this coin is more encouraging. If we can properly define the problem, then we are more than halfway to solving it. We therefore need to explore and develop a fuller understanding of how best to present the information which is needed to build that deeper, critical understanding of what we see and hear.

If we design the transmission medium to carry only superficial signals at the expense of the finer inflections then we have failed. Just like the $200 CD player that muffles the musicians shuffling their feet we will have removed the clues that enable our children to embark on their journey of discovery as they first start to build that initially fragile and flimsy mental mesh. Without the presence of the finer audio-visual nuances and subtleties, which will nourish their deeper understanding, their mental cognitive reference grid—their nascent critical faculty— will remain in tatters. Our children will remain at first base.

This is no inheritance to pass on to our heirs.

So, let us reexamine digital transmission while keeping the idea of cognitive mesh building in the back of our mind.

Let's recap.

We enjoy two sensory mechanisms. The first is the instant or "at first glance" reaction and the second is the much slower buildup of a more critical eye or a more discerning ear.

Nowadays it is relatively easy for the digital communications engineer to produce pictures and sound that "at first glance" appear satisfactory but that later on appear to be "wanting." If we do not know what is missing this can lead either to an over-provision or to an inappropriate presentation of the information.

It could be that the missing ingredient is very easily added—just as an extra half spoonful of sugar can turn a cup of coffee from a bitter taste to something much sweeter. Or it could be that the missing ingredient is a complex mix of lots of different elements.

The "ingredients" and "recipe" analogies are drawn of course, not from electronics but from that much earlier and probably more important science—that of cooking. The following analogy—based on the baking of a cake—sheds some useful light on how digital communications electronics and a CD player really work. We shall pick up on the ingredients and recipes later in the chapter.

## A BYTE OF CAKE

Although the microphones in the UK parliament pick up every word that is spoken not a syllable of these discussions is transcribed onto paper. That task is assigned to the official band of Hansard stenographers who record and publish the sanitized version of what is said. Their work regime is tough—a rolling schedule of ten minutes of transcription followed by ten minutes off. Stopping the parliamentary proceedings is not allowed so any possible mistakes or omissions must be taken up and corrected in the ten minute breaks between transcriptions.

At the end of one of the broadcasting debates I fell into conversation with one of the Hansard stenographers. "I am confused," he said "Are the MPs (Members of Parliament) really serious when they talk about bites of cake?"

"They are quite serious." I replied. "The cake analogy may be their only nontechnical lifeline to the meaning of 'digital.' And can you spell it byte not bite?"

I was of course delighted that the MPs had taken up analog and digital cake with such gusto. It allowed them to lift the veil and peep at the mysteries of analog and digital communications without the need to digest pages of unfamiliar mathematics.

## AM and FM analog transmission

Imagine a bakery that makes large, delicious, fancy iced cakes with lots of soft piped cream on top. Not only is every care taken in the production of each cake but every care is also taken in transporting each cake from the bakery to our local store.

Unfortunately, accidents can and do occur. However much care is taken by everyone involved, the cake will inevitably suffer some damage, perhaps gaining one or two minor dents in its journey from the bakery to the store and perhaps then gaining a number of more obvious dents on the way from the store to home. Once the "accident" has occurred there is no way to put it right.

Radio or phonograph transmission from the concert hall to our homes is just the same. Even though every care is exercised in the choice of landlines from the concert hall to the radio transmitter or in the "cutting" of the master record some slight imperfections will inevitably creep in.

The journey from the transmitter to our radio set or from the phonograph album to our loudspeaker is even more hazardous. "Static" may affect our radio reception and dust may collect on our albums. Our ears will recognize and perhaps still appreciate the music even though it is now but a faint imitation of what left the concert hall. We would probably fervently wish we were there, in person, to listen to the concert. By analogy (sic) the cake has been stripped of nearly all its topping but we still recognize it as, say, blueberry muffin or cherry cake.

## Digital transmission

Let's get back to the bakery. The new owners of the bakery have decided that dents must be abolished forever so they have invested in a new digital bakery with new quality control methods.

There is a new packing process.

Once the bakery has completed the cake it is chopped up into a million or more pieces. Every piece of cake is cut to the same size. Some pieces will contain plain cake, some will contain topping and some pieces will contain a little of both. Each piece is carefully and individually wrapped to make sure that it will come to no harm in transit.

From the moment they are safely packed for their journey from the bakery these myriad pieces of cake are not unwrapped until we get them home from the store. It is then our task to stick them all back together to make the original cake. We must sit down and start rebuilding the cake—just like doing a giant jigsaw puzzle.

If, by chance, some of the pieces have been lost or damaged it is not too difficult to guess whether a replacement piece should be a cake piece or a topping piece. Help is at hand, the parcel contains some extra "joker" pieces which we can use.

It does not take long to realize that we also need a very steady hand if we are to glue the pieces together properly. Any uncertainty or irregularity in the assembly process will certainly show up in the outward appearance of the reconstructed piped cream though it matters less about the appearance of the blueberries inside the cake.

> The compact disc uses the same cake-wrapping principle. It contains millions of chopped pieces of the music—all cut to a standard size—just like the standard-sized distribution of pieces of a wedding cake that will be sent out later to the invited guests.

The domestic compact disc player must now undertake two quite difficult tasks:

First          It must repair any transit damage. Its error conceal-
               ment circuits must spot any damaged piece of music

and substitute a replacement piece. This is usually done by taking a duplicate copy of its error-free neighbor. This technique works fairly well for pieces that are all-cake or all-cream but less well for mixed pieces.

Second    The CD player must then assemble and glue the pieces back together. Unsteady joins will produce a very ragged sound whereas slightly unsteady joins will fail to recapture the subtle texture of the topping—instead producing that "bright" sound which is a characteristic of cheap CD players.

Which is better: analog or digital?

The answer to this question depends on whether you believe home constructed (home-baked) or store-bought cake is better. This, in turn, depends on how good a cook you really are.

Analog    The store-bought cake (the phonograph album) is ready to eat and is well presented—though it may have suffered a few dents in its journey home.

Digital   The home-baked cake (the compact disc) needs a lot of construction and cooking skill if it is to out-do the store-bought cake. Without that degree of skill the more subtle flavours will be lost.

For these reasons Hi-Fi buffs are currently torn between the two technologies. The 5-inch (12-cm) compact disc is clearly a more convenient format but we must take care in the reconstruction of the music. It requires a very steady hand if its quality is to exceed that of the old-fashioned phonograph album. In time, however, it seems probable that the necessary home-cooking skills will be successfully embedded in the CD player and the quality of reproduction will then outstrip the vinyl album.

This must appear a strange and unexpected result as we know that both the modern 12-inch long playing album and the 5-inch compact disc have, at some time, probably been processed in the studio in a digital form. It just serves to show how a slight shift in the form of presentation can make a great deal of difference to our perceptions. The next chapter will explore this perceptual difference in more detail.

**Double standards for audio**

In digital terminology the compact disc offers a resolution of 16 bits. In other words it allocates the volume of the sound a series of distinct values somewhere in the range of plus or minus 30,000 or so units. When 16-bit digital recording was first introduced this wide range was welcomed enthusiastically as meeting all foreseen requirements. After all, who can hear the apparently imperceptible midrange difference between, say, 10,000 and 10,001 units of volume?

Some can. Since the introduction of CDs the newly retrained human ear has caught up with 16-bit technology, has noted its defects and now demands even higher resolution.

Recording studios are now investing in 20-bit recorders which offer a resolution range of plus and minus 500,000 smaller units of sound volume. Some manufacturers have even introduced 24-bit recorders for recording live events where there is less control of volume levels.

Surely no one can spot the midrange difference between 100,000 and 100,001 smaller units of volume. Or can they?

Let us trace the sound path through the digital studio. The first step is to capture the artists' live sounds through a number of microphones. This is done by recording each sound on separate tracks of a multichannel 20-bit studio recorder. Once this multitrack recording is safely "in the can" the artists can return home and the job of the mixing engineer can then begin.

Whereas the recording of the separate microphones onto the separate tracks of the studio digital tape recorder is considered to be "squeaky clean," it is most unfortunate that the subsequent digital mixing process nearly always causes distortion.

In the post-production process the various microphone "tracks" will be mixed and re-mixed together in different proportions. This is achieved through banks of digital volume controls or "faders" that each introduce a small amount of distortion into the last few bits of each digital sound sample.

How does the distortion creep in?

In the digital domain the straightforward mathematical processes of adding and subtracting do not cause errors or distortions. However, the more complicated processes of multiplication and division will nearly always cause some distortion due to so called "rounding errors."

Let us try an example. Reducing the volume of sound from a singer's microphone to, say, a third of its initially recorded digital value will cause distortion because, quite simply, the digital domain does not like fractions.

A more familiar example of this dislike of fractions is our pocket calculator's inability to calculate some numbers accurately. In the notation of our schooldays' vulgar fractions a ratio of one-third is always shown on the calculator display as 0.333 etc. but never quite ⅓. More worrying is that we know that a digital third (0.333) multiplied by three should be one but the calculator will display 0.999. The difference between 0.999 and 1.000 is 0.001 or, in percentage terms, a distortion of 0.1 percent.

Can we can get around this problem? Yes, but only in part. We can reduce (but never quite eliminate) this rounding error distortion by just adding more "places of decimals" to the calculator display. After all, the difference between 0.99999999 and 1.00000000 is pretty small!

The same trick can be applied to the recorded music.

As the distortion or "rounding error" will only crop up in the last or least significant few bits of each sound sample, by far the easiest solution to the problem of mixer distortion is to just extend the sound sample by the addition of an extra few "sacrificial" bits. These can then be discarded later on. Hence the need for 20-bit studio recorders which allow the audio recording engineer to mix and remix the music without incurring ever increasing and excessive levels of distortion.

It is only in the final stage of production that the 20-bit samples are reduced to the 16-bit format which is needed for the compact disc master recording—though this process also needs some care in its implementation.

Simply slicing off (and hence ignoring) the last 4 bits of the 20-bit sound samples in order to meet the 16-bit compact disc format has, in practice, produced a noticeably "gritty" sound. The audio engineer must instead try and recapture a nuance of the information that was contained in the discarded last 4 bits by the process of intelligently "dithering" the 16th or least significant bit on his CD master recording.

For the purist this "dithering" process does not cure the rounding error problem, it merely substitutes a less noticeable form of distortion by blunting the rough edges of the 20 to 16 bit slice. The purist would much prefer to listen to the full 20 bits of the recording and questions the need to slice off the finest 4 bits. In time his wish may be granted. It may prove possible to trade off bits for playing time so that 60-minute 20-bit CDs may one day become as popular as today's 74-minute 16-bit CDs. At present, however, the domestic CD player is designed to accept music only in a 16-bit format and would be confused by any change.

The audio recording engineer would also appreciate some flexibility at the other end of the scale. He prefers to leave the higher echelons of the volume levels untouched, just in case an especially loud sound were to come along unexpectedly. Thus the CD recording is unlikely to use up more than 13 of the 16 bits available, leaving a 3 or 4 bit safety margin for what audio engineers describe as "headroom."

### Double standards for video

Just like his audio colleagues, the TV engineer is also seeking some headroom in manipulating the TV picture. The same apparent over-provision of digital bits is increasingly common in the very specialized area of television picture post-production. Although quite satisfactory television pictures are sent every day in an 8-bit format, it is now thought preferable to massage the TV signal through special effects units in an enhanced 10-bit format.

When the TV picture emerges from the special effects it must be reduced from a 10- to an 8-bit format. It is possible to use the same "dither" techniques as was used in the sound processing described above to convey a hint of the finer information that was previously available in the 10-bit format.

TV broadcast companies always demand the highest fidelity in the incoming pictures—via the so called *contribution links*—as these incoming pictures may need to be massaged and mixed together in the studio. It is very important that these "raw" pictures are of the best possible quality: the pictures cannot get any better in their passage through the studio, they can only get worse. Once the final pictures have been created, however, they may be distributed at somewhat lesser quality or, more accurately, with somewhat less headroom—via the so called *distribution links*—to the network of TV transmitters.

In the past broadcasters have specified and obtained high "contribution quality" cable and microwave radio TV links from phone companies and other telecommunications service suppliers. At present these links are still mostly analog. For convenience the broadcasters have used the same quality of links for the subsequent distribution of their programs. Thus the quality of the TV picture is well preserved right up to the terrestrial high-power TV transmitters. Future developments in digital television transmission and possible changes in the tariffing of these circuits may make it worthwhile to discriminate between the required qualities of contribution and distribution links.

There are thus two equally important sets of digital standards which are used by broadcasters:

First      The contribution quality links that are engineered to offer a transmission quality of as much as 20-bit audio and 10-bit video

Second     The distribution quality links that offer lesser transmission qualities of 16-bit audio and 8-bit video.

The first set of standards, for the contribution links, are used almost entirely within broadcast organizations whereas the second set of standards are those that will reach out to us at home. It is therefore always important first to establish with the broadcaster which set of standards he intends to apply, especially when judging the quality of the digital compression techniques that lie at the heart of digital high definition television.

### More double standards

#### 1. The personal computer

The sort of word processing (WP) software that we use in our personal computers also enjoys a double standard in the storing of its documents or "files." In computer language the revisable WP format equates to the production and post-production stages of making radio and TV programs whereas the final WP format is, not surprisingly, the same as the finished radio or TV program.

The revisable word processing format is quite complicated for it includes information about tabs, indents, soft hyphens, right justification and all the minutiae of setting and resetting the text on the page. It leads to sizes of computer files that are often much bigger than were first expected. This extra housekeeping information is essential whether a document is either to be networked between a number of writers and editors, each of whom may wish to make some corrections or additions to the original script or, more simply, whether the writer would like to add some further amendments to his own script later on.

On the other hand the final word processing format is relatively straightforward. Stripped of the housekeeping information, the "file" is much smaller in size than the revisable format as it is, rather hopefully, assumed that no one will need to change it later on. In practice our personal computer will always play safe and choose to store the larger, revisable format of the WP file. Difficulties only arise when one's colleagues use a different make or version of a word processing package. Even then the basic text is unlikely to cause a problem—it is more likely that the extra proprietary housekeeping information will be mis-interpreted by the other WP software.

### 2. The fax machine

Although a fixed resolution digital system can often appear to be beyond reproach yet sometimes it can seem to be inadequate. A good example of this apparent double standard is the ubiquitous digital Group Three (G3) facsimile machine—the fax. The fax can sometimes offer us excellent results though the everyday quality is normally not quite as good.

- The best quality can be seen in the header at the top of the received page or in the fax "journal printout" where the fax machine is, in both cases, able to choose for itself where to place the individual dots of the written script. The intrinsic typographic quality of the script is, in reality, quite appalling but, despite that limitation, it is amazingly clear and easy to read.

- The middle quality is usually seen when a little care has been taken in aligning the typed message on the transmission sheet with the paper guides of the fax machine. The horizontal scanning mechanism within the fax machine can now match up with the horizontal lines of the typed message on the transmission sheet so that the received page is quite easy to read.

- The worst quality is occasionally seen when the typed message has been allowed to enter the fax machine at a slight angle. The received fax is clear in some parts but fuzzy in others. What should have been straight horizontal lines appear to have been replaced with short line segments that resemble sets of badly drawn shallow steps.

In recent years the purchase of a personal computer "fax card" has offered fax transmission straight from the computer stored document to a remote fax machine. In theory such a "paperless" approach allows the computer to choose where to place the dots on the remote page and hence should ensure very high quality results. In practice, however, the fax card communications software is still a little cumbersome to use and the fax card owners often find it more convenient to pop a piece of paper in a conventional fax machine instead.

## THE SUPER-TRUTH IN THE SNOW

The analog cake in the previous section needed a lot of care in its packaging in order to keep it from harm in its journey from the bakery to home. Even then it might still suffer the occasional "accident" and thereby pick up the occasional dent. Despite the advantages of the digital cake alternative there is a downside. The digital cake does need far more packaging and wrapping paper for its journey home for we know that each piece must be wrapped separately.

Properly packed digital cake is far more bulky than analog cake but it can be transported anywhere with no dents at all.

This super-truth was discovered by William Shannon some 50 years ago. He found that it was possible to design wrapping materials—he called them "coding schemes"—that could carry digital cake from one end of our universe to the other without any damage at all.

Until then it was generally believed that the further the cake was carried the more damage it would inevitably suffer. Everyone knew, did they not, that radio signals would gradually fade out the further away one got from the transmitter. Telephone lines would always crackle and hiss on long distance calls. The only way to alleviate these problems was to build more powerful radio transmitters and install fatter telephone cables.

Shannon changed all that.

He showed that, with a little care, the initial state of grace— our initially perfect cake—could be transported and preserved indefinitely. A subsequent decline from grace, whether swift or slow, was no longer inevitable. Simple packing precautions in the digital bakery could preserve our cake inviolate.

There was, of course, a downside.

If we ease up on the packing precautions the cake will turn to crumbs and will be lost—it is then beyond redemption.

In practice it is not difficult to clear these hurdles and provide error-free digital transmission though there are some other trade-offs to be considered. First, however, we must find out how much margin is needed to clear these hurdles.

To discover this let us imagine we are now trying to order our cake from across a crowded restaurant.

- If we are ordering the cake from the analog waiter we might ask for, say, a cherry and walnut cake with various exotic toppings. We know that the distant order-taker might easily mis-hear our order unless the restaurant fell particularly quiet at that moment.
- If we were ordering our cake from the digital waiter we could rattle off the individual elements of the cake as "Cake-cake-cake-topping-topping-cake-cake," etc. above the general hub-bub in the restaurant and still get exactly what we want.

Two points emerge from this example.

1. We will need to rattle off the digital cake specification at very high speed if we are not to be there all day.
2. The staccato "cake-cake-topping" sequence will punch through the general hubbub of the restaurant with ease whereas the finer nuances of the analog cake's specification might be muffled by even the gentlest sound of tea cups.

It is this need for a very high-speed staccato delivery of the digital order for our cake that is the fly in the ointment of digital transmission. Later in this chapter we will discover how the rate can be reduced many times over yet still retain the error-free advantage of digital over analog transmission.

### Quiet please!

The analog restaurant waiter must be often tempted to utter these words when trying to listen to an order. We have always referred to unwanted sounds as "noise," the same term is also used for unwanted electrical signals.

"Electrical noise" includes the interference or "snow" that we occasionally see on a poor television picture and the hiss that we can sometimes hear on a long distance telephone connection.

Carefully controlled listening tests have shown that people with normal hearing can correctly identify about half of a set of random English test words which are delivered to them via headphones at a strength 15 deciBels (dB) above the background noise level. In nontechnical everyday language this signal/noise ratio of 15 dB means that the *average* microphone voltage level, which occurs as the test words are spoken, is about five times greater than the *average* background noise level that would be picked up by the microphone in the absence of any speech.

At first sight it seems difficult to understand why such an apparently low background noise level should cause such havoc to the intelligibility of the words. The answer lies in how we have defined the "average" value of the noise. For example, taken over the course of the year, the average level of our local river may be a few feet below the top of the river bank but this knowledge provides little comfort when the river overflows for a few hours during a winter storm. Table 1-1 shows that short-term fluctuations of typical background noise voltage level will:

**Table 1-1**    Statistical Short-Term Variations in Noise Voltage
Compared to the Long-Term Average Value. The
Final Column Expresses These Variations in the Db
(deciBel) Logarithmic Form.

| | | |
|---|---|---|
| often | reach twice the average value | 6 dB |
| sometimes | reach three times the average value | 10 dB |
| occasionally | reach five times the average value | 15 dB |
| very rarely | reach ten times the average value | 20 dB |
| almost never | reach twenty times the average value | 25 dB |
| never | reach thirty times the average value | 30 dB |

The intelligibility of English speech is a complicated subject—the longer vowels sounds can punch through a noisy environment but the shorter consonants fall easy prey to just a little noise. It is a lucky man who can successfully tell the difference between a "c" and a "g" or a "p" and a "b" on a poor quality telephone call.

Let's check the table.

At an average signal/noise level of 15 dB the noise will occasionally be as loud as the test words and this will obscure or "mask" what we are trying to hear.

If we rerun the listening tests at higher signal/noise ratios we find that the intelligibility of the random test words approaches 100 percent as the signal/noise ratio rises to 30 dB. It is worth noting that these are unusually tough tests as, unlike normal speech, the random choice of each test word bears no relation to and hence provides no clues to its neighbor.

The results, however, only relate to English test words. Some Oriental languages, such as Chinese, convey meaning by relying more heavily on the inflection of rugged vowel sounds than on the use of fragile consonants. As a virtue of their language the Chinese may thus be able to communicate in more noisy surroundings than their occidental cousins.

These signal/noise figures are indeed correct but they do not cover one important eventuality—the accident.

Every now and then the general background hubbub in the restaurant may be entirely swamped by an extremely loud crash or series of crashes from the direction of the kitchen as a pile of dinner plates hits the floor. Conversation will temporarily stop as everyone regathers their wits.

It is this totally unexpected burst of noise, far louder than either the background hubbub or our own conversation, which can bring communications to a halt. We will return to this concept of burst noise later on.

In order to get a better feel of the dB levels let us first examine a few more familiar examples of consumer electronics.

### Audio systems

- An audio cassette tape recorder or the sound track of a standard (non Hi-Fi) video cassette recorder (VCR) will offer a signal/noise performance of about 42–45 dB. This means that the signal voltage is 150 times stronger than the noise. This may improve to 48–52 dB (300 times above noise) if a proprietary form of noise reduction (such as "Dolby") is switched in.
- Long playing albums, compact discs, good quality VHF FM radio reception and TV stereo sound offer much better signal/noise ratios in the range 60–80 dB (1,000 to 10,000 times above the noise).

If intelligibility was to be the only criterion we could get by, at a pinch, with communication systems that offered no more than a 30 dB signal/noise ratio. Old-fashioned AM radio is a good example of such a system. As we know, however, there is little margin or "headroom" for speaking quietly or for being able to listen easily to the quieter passages in the music.

Our telephone system is somewhat better than AM radio. It offers a signal/noise ratio of just over 40 dB (100 times above noise) on international calls and more than 50 dB (300 times) on local calls. This performance is considered quite acceptable for "commercial speech" but is still barely adequate for the transmission of music as we know from the unpleasant hiss on the soundtrack of many older rented video cassettes.

## TV systems

Viewing tests suggest that a television picture is quite satisfactory at a signal/noise ratio of more than 50 dB (300 times above noise) whereas "picture snow" becomes only too apparent at signal/noise ratios of less than 40 dB. Cable TV system designers aim to provide customers with television pictures that enjoy signal/noise ratios of better than about 45–48 dB. At this intermediate signal/noise ratio the snow is no longer visible but the picture does not seem quite as "solid" as that from a 50 dB s/n source.

The slightly lower figure of 43 dB (150 times above noise) serves to define the normally accepted "snow limited" range of a terrestrial TV transmitter.

A comparison of these figures shows that the human ear appears to be considerably more sensitive to aural noise than the human eye is sensitive to "picture snow." After a short time our ears will notice noise some 70 or more dB below the audio signal unlike our eyes which cannot "see" the snow if it is more than 50 dB below the video signal (3,000 and 300 times respectively).

Let us return to ordering our cake in the restaurant.

The analog cake waiter will need to hear his customer at a level at least 30 dB and preferably more than 40 dB above the background hubbub in the restaurant if he is to get the order correct.

What of the digital cake waiter?

He must lead a relatively humdrum and mundane life as people only ever address him with cries of "cake" or "topping."

He is able to recognize no other words.

Because of these strict limits on his vocabulary he is able to detect cries of "cake" and "topping" over vast distances, just as we are said to be able to hear our own name immediately when whispered from afar at a cocktail party.

Let's switch from audio to video communications. If the digital waiter was really up-to-date we could avoid shouting across the restaurant and could try televisual communication instead. The "cake-cake-topping" script could be written down in the form of the striped bar code that now appears on every item of groceries in the food store. A black stripe could represent "cake"

**Figure 1-2** Calling the digital waiter.

and a white stripe would represent "topping" so that "cake-cake-topping" becomes the striped sequence "black-black-white."

Imagine holding up such a black and white bar code (Figure 1-2) in front of a nearby television camera so that the close-up completely fills the screen of the waiter's distant television set.

Finally, imagine how incredibly snowy or grainy the picture must now need to become before the digital waiter would fail to recognize the picture as a bar code and fail to interpret correctly the black and white stripes as our request for cake.

We may have experienced similar black and white striped pictures when setting up a new video recorder. At the back of the VCR there is a switch marked "Tune" When this is depressed a series of black and white vertical stripes will take the place of the normal playback picture from the video cassette. The black and white pattern is intended to make tuning a free channel on the TV to the output of the VCR very much easier. The pattern is so distinctive that it can still be recognized on screen even when the antenna lead between the VCR and the TV set is temporarily disconnected and held an inch or so away from the TV socket.

Theory shows that we can always correctly distinguish the black and white stripes at a signal/noise ratio of no more than 20–25 dB (10 to 20 times above noise) though our success rate will fall off rapidly below this.

At less than a 12–14 dB signal/noise ratio our success rate plummets to zero—the digital packing material has fractured and the pieces of cake have turned to crumbs.

This ability to work perfectly well right down in the "picture snow" is the second reason why digital communication is such a godsend.

Whereas the analog signal/noise ratio of a telephone or television signal can never be allowed to fall below 48–50 dB on any of the many "hops" of its journey from one end of the country to another, the digital signal can be safely allowed to carry on getting steadily worse and worse for mile after mile beyond the range of the analog signal. Providing the digital signal is caught and re-formed before its signal/noise ratio falls below 20–25 dB then all is well—just like catching that expensive antique vase a split second before it hits the floor.

### Digital profitability

Until now the main beneficiaries of this digital transmission technology have been the telephone companies. Long ago they acquired the monopoly for moving telephone and television signals the many miles around the country and around the world on our behalf.

In turn the telcos have invested considerable sums of money into developing digital technology. Their research departments have been well funded and their research staffs have been encouraged both to cooperate and compete with their peers in similar organizations throughout the world.

It would be difficult to appreciate modern digital television distribution unless we first understand the bedrock of digital expertise which has been created on our behalf by the telcos. This is described in the following section which is entitled "Our Telephone Inheritance." However, it would be naive to expect that this technical investment was entirely altruistic—research efforts remain directly focused on the benefits they can produce for the business. The telcos have therefore applied digital transmission technology to their everyday business in the following way.

Unlike the earlier analog signals, which needed to be boosted every seven or eight miles, digital signals can easily travel the typically 30 or so miles between major telephone exchanges without mishap. The telcos have thus been able to make useful savings by scrapping the intermediate analog booster stations which had always been thought of as something of a security risk.

But why stop there? Why not also scrap the local town's

telephone exchanges and instead extend everyone's telephone line the 30 or so miles to a major switching point or "hub"?

In furtherance of this policy most local exchanges are being replaced with "dumb" digital line extenders and, consequently, many telephone numbers are being changed to reflect the new hubbing arrangements.

Shannon's ideas of digital transmission have fueled a revolution in telecommunications. We can now telephone friends in Europe, Hong Kong and Hawaii and often enjoy the same clarity of speech as in a local call. However, telecommunications tariffing is still heavily based on the pre-Shannon idea that longer distances imply fatter phone lines and hence higher costs.

This is no longer true. The cost of international connection is now relatively fixed, lying principally in the near and far end switching equipment rather than in the interconnecting loss-free satellite and fiber optic cable links.

## OUR TELEPHONE INHERITANCE

It is easy to forget that, until recently, only the telephone companies were permitted to move the broadcasters' signals around the country—the telco monopoly forbade anyone else to do it for themselves. This monopoly also effectively extended to technical matters. After all, what business was it of the customer to exert technical influence on the development of the telephone network?

It is therefore important for us to understand the considerable intrinsic value of this technical inheritance. The products of many decades of technical monopoly are set to exert a strong influence on what we might easily attempt in the future.

The fly in the ointment of digital transmission is that the signaling rate—the speed of the "cake-cake-topping" sequence—needs to be very fast. To support the human voice we need digital signaling rates of tens or hundreds of thousands of digital elements per second. Television pictures need even faster signaling rates of tens or hundreds of million elements per second.

The electronic circuits and cables that carry these very high speed signals are not particularly difficult to design or make, the principal difficulty lies in the fact that they are just different from the analog circuits and cables that went before.

This incompatibility can be tackled in two ways:

- The first approach is to lay a completely new grid of digit-friendly cables—a national information highway.
- The second approach is to somehow "doctor" the digital signals to make them pass more easily down ordinary, old-fashioned, analog-friendly cables.

In practice, of course, it is worth backing both horses.

Although it has proved an expensive investment the telephone companies have been able to install a comprehensive mesh of brand new fiber optic digital cables which duplicate the earlier analog routes. In the near term this considerable investment is containable as they need to interconnect no more than a thousand or so major telephone exchanges throughout the country. This is not cheap to do but the cost and benefits of each interconnection can be accurately determined and budgeted in advance. New telephone exchanges rarely appear overnight as they are normally part of well understood rolling 5 or 10 year business development plans. It is therefore relatively easy to arrange to connect each exchange, on time, to the evolving mesh of digital cables.

On the other hand, the telephone companies' customers are far more fickle.

We expect to and, indeed, do change both our domestic and business plans far more frequently than the comfortable ten year development time scale which telcos have been able to enjoy in the past.

In the eyes of the phone company some of their customers are not only easier to deal with but appear to offer more profitability than others.

The operator of a local petrochemical refinery is unlikely to pack up and move the production facility to another state overnight whereas a retailer of fast moving consumer goods (FMCG) might easily be tempted to move by the prospect of cheaper warehouse accommodation either a few or a few hundred miles away.

For the phone company it thus seems far more worth while to lay a new fiber optics cable from the nearest telephone exchange to the refinery than to install a similar link to the warehouse complex.

In practice, however, each customer's business interests will range well beyond the confines of the refinery or the warehouse.

For example, the petrochemical business may operate not only a refinery in Texas but also a chain of warehouses that stretch from New England to California. In addition let us not forget his franchised resellers who are also scattered all over the country and who may also need high speed connections to the central warehouse computer system.

The ideal engineering solution would be, of course, to lay fresh digit-friendly cables to every business and domestic customer in the country—but at what cost and why?

Providing high capacity links between a thousand or so telephone exchanges can make immediate economic sense but recabling the last few miles to each of more than 90 million homes in the United States requires a strong will and very fat pocketbooks indeed.

This recabling is immensely expensive and offers no clearly guaranteed return on the investment.

The telephone companies therefore have derived two plans:

Plan A      The new digit-friendly fiber optic customer cables would appear to offer economic viability if it is possible to carry television programs as well as conventional telephone services over the new cable.

Plan B      The old analog cables might just be able to carry some enhanced digital telephone services the last few miles to the customer if the digital signals were "doctored." Business users seem willing to pay for video conferencing links but the cost of the "doctoring" process may be too high for the transmission of better quality television pictures to the more distant residential market.

Both plans are in trouble.

Plan B—otherwise called the single line Integrated Services Digital Network—or ISDN for short—requires great care in its implementation. The old analog cables were never intended to carry digital signals and so need individual "tweaking" if they are to be made to work properly over distances of more than a mile or so. This need for individual attention on each old-fashioned line offers the telephone company no economies of scale. It therefore remains expensive to implement on a wide scale and this will probably limit the ISDN service to business rather than residential applications.

In the long run it would clearly make better sense to install new cables which avoid the need to pay individual attention to each line.

This leads us back to Plan A.

Plan A has proved unpopular with both the cable TV companies and the politicians. If the telephone company is permitted to provide both telephone and television services through the same cable, it will enjoy a very strong monopolistic position—a position that many politicians would prefer to see diminished rather than enhanced. The telephone companies may therefore come to accept that their only way forward is to strike deals with the competing neighborhood cable TV company.

Plan A is founded on an implicit assumption which has yet to be thoroughly tested or proved. The unstated assumption is that fiber optic digital cable will offer us a better means of delivering television services to our homes.

The final chapter, which is entitled "Simply Irresistible," will reexamine this assumption.

## Hierarchies

Many years ago the telephone companies agreed on a set of digital transmission standards that were intended to carry speech signals around the world.

It was first agreed that adequate speech quality could be achieved if the speech waveform which was picked up by the microphone in the telephone handset was sampled 8,000 times a second. It was also agreed that each sample should be resolved to an accuracy of about plus and minus 120 units of volume—i.e., a resolution of 8 bits. There was disagreement between America and Europe as to how the volume units might best be "doctored" in order to improve the fidelity of the sound at low volume: the Americans defined their chosen method as "μ law encoding" against the Europeans' rival "A law encoding."

Translation between the two "laws" on international telephone calls is entirely straightforward but is subject to the same sort of rounding error distortion which we discussed earlier in the section entitled "Double standards for audio" on page 14.

What was most important about the agreement was that the basic bit rate of both systems was to be the same everywhere in the world—8 bits repeated 8,000 times a second—that is 64,000 bits per second or 64 kb/s for short.

We know that, in order to enjoy economies of scale, individual telephone calls are first gathered together in telephone exchanges and then routed, in bundles, most of the way to their destinations. Near their destination another exchange or switching point fans out the calls to the individual recipients.

The old-fashioned analog trunk lines were designed to bundle the calls together, first in sets of twelve and then in supersets of five of the smaller bundles—60 channels in total. The telephone calls were stacked up in a heavy-duty trunk cable at different frequencies just as various radio stations are spread across the VHF/FM dial of our radio set or as various TV stations all travel down the same antenna download cable from the rooftop TV antenna to our TV set. Sometimes, on busy routes, the 60 channel bundles would be further stacked up in twos or fours to establish 120 and 240 channels on a single cable.

Telephone companies worldwide became quite used to handling channel numbers such as 60, 120, 240 etc. both on their main trunk circuits and on their under-sea international cables.

The question now arose:

Should the new digital circuits offer the same step size as before or was there an opportunity to make improvements?

### Level one

Once again, there was some disagreement as to the best course of action. The Americans chose an initial grouping of 24 channels whereas the Europeans chose 30 channels. This resulted in bit rates of about 1.5 million bits per second (1.5 Mb/s) for the Americans and about 2 Mb/s for the Europeans. However, the new method of stacking the channels together was quite different from the "multiple radio station" or "Frequency Division Multiplex" (FDM) method which had been used for stacking up analog telephone channels.

The new digital stacking method was called "Time Division Multiplex" or TDM for short and predates by many years the now familiar format of the Cable News Network's Headline News TV channel.

In CNN's Headline News service we are usually offered news headlines from the top of the hour to five past the hour. This is then followed by a number of special "spot items." International weather might be scheduled at 14 to 16 minutes past the hour and sports news might occur a few minutes after that. However, if we accidentally turn on the TV for headline news a few minutes late then we miss the headlines and must wait until it comes round again some 30 or 60 minutes later.

Thus the digital telephone transmission format stacks all the voice channels behind one another on one super-channel rather than assigning individual radio frequencies to each voice channel.

This is much more efficient than the old analog method but falls down completely if synchronization is lost—just as if we turn on the TV too late and miss the headlines at the top of the hour.

The 1.5 and 2 Mb/s standards were only the beginning.

### Level two

Both camps agreed that the next layer in the hierarchy would comprise four initial bundles so that the American second level digital hierarchy contained 96 channels and the European hierachy contained 120 channels at just over 6 and 8 Mb/s respectively. In practice neither the Americans nor the Europeans have taken much interest in this level, they prefer to jump straight from level one to level three.

### Level three

There was disagreement at the third level. The Americans bundled seven of the 6 Mb/s "tributaries" together to form a bit stream of just under 45 Mb/s which could support 672 voice channels. The Europeans bundled four 8 Mb/s tributaries together to form a 34 Mb/s bit stream of 480 voice channels.

### Level four

At the fourth level, 140 Mb/s, the current *de facto* standard is derived from the European hierarchy. AT&T also proposed a 140/280 Mb/s standard which would combine three or six 45 Mb/s tributaries but the international marketplace has instead adopted the almost identical European 140 Mb/s standard based on four 34 Mb/s tributaries. At 140 Mb/s the AT&T proposed standard would have supported 2,016 voice channels whereas the European standard supports 1,920 voice channels.

Higher levels in the digital hierarchy are still appearing. New international sub-sea fiber optic cables are now designed to support even higher rates. These use multiple 140 Mb/s tributaries which are combined together to form gross speeds of up to 2,500 Mb/s and beyond.

It seems strange, does it not, that the marketplace preferred the European 140 Mb/s standard when the homegrown AT&T proposed standard could squeeze in nearly 100 more voice channels—2,016 versus 1,920 channels.

What made the telephone companies back off?

Both the Europeans and the Americans have come to realize that they have overdone it. 1,920 voice channels, let alone 2,016 voice channels are too many to deal with comfortably.

By example, imagine taking a touring vacation in a car with a small trailer in tow. We are given the choice of an 18-foot or a 14-foot trailer but choose the smaller one. Whichever trailer we choose it is inevitable that we will still need to take just as many ancillary items for the journey. So, after a great deal of muttering, we may finally congratulate ourselves on having packed everything into the smaller, 14-foot trailer.

But have we really done ourselves a favor or not?

Every time we stop for the night and need to find those few

important overnight items we must first unload nearly everything in the trailer in order to get at the cooking utensils or the
knives and forks. Would it not have been a better plan to have
taken the larger 18-foot trailer instead and then been able to put
our hands on any small item immediately?

The telephone company engineers have learned the same
lesson. Trying to unbundle a few voice channels from the high
speed multiplex has proved to be heartbreakingly difficult. The
timing has to be so precise and each of the hierarchical layers
must be unpicked—or demultiplexed—in turn, one layer at a
time.

They have learned that there are no more Brownie points to
be won for stacking extra channels in unexpected places of the
high-speed multiplex. It is much better to build a much simpler
stacking system which has far more built-in clues to indicate
where everything has been stacked.

In order to achieve this simplification it has been agreed to
raise the overall bit rate by about 10 per cent, to 50 Mb/s from 45
Mb/s and to 155 Mb/s from 140 Mb/s. This is the essence of the
much touted Synchronous Digital Hierarchy (SDH) (also referred
to as Sonet when used on fiber optic cable).

It is just as though the CNN announcer had been told to
deliver the news item just a little faster in order to make room for
a one second pause before and after the advertisement break.
European viewers are both used to and welcome such visual
punctuation marks whereas American viewers have become accustomed to seeing no perceptible break between the news and
the "messages."

The older 140 Mb/s and the newer 155 Mb/s services will
coexist together for many years. Even the older Frequency Division Multiplex method that was used to stack up analog channels
has undergone a metamorphosis and has taken on a new lease of
life.

Developments now allow us to transmit more than one "CNN
Headline News service" by simply using many different lights of
different colors down each fiber optic strand. Thus the "CNN"
multiplex can remain on, say, the blue channel while a completely

independent "C-Span" multiplex is transmitted down the same fiber on the red channel.

This explosion in the carrying capacity of the cables can only be achieved on new fiber optic cable—the telephone companies' Plan A.

But it is not the only game in town.

## COMPRESSED DIGITAL TRANSMISSION

The previous section showed how the established telecommunications networks throughout the world have settled on certain favorite digital transmission speeds. They are all based on a metronome that "ticks" 8,000 times a second and use a basic transmission building block of 64 kb/s.

In Europe these 64 kb/s units are successively bundled into 2, 34 and 140 Mb/s "bearer" circuits whereas in America the 64 kb/s units are bundled slightly differently into 1.5, 45 and, occasionally, 140 Mb/s bearers. When the digital hierarchies were first designed it was thought that these high speed circuits would be primarily for the telephone companies' own use.

In the event, however, the high speed circuits have also proved of great value to many of the telephone companies' larger customers.

In the UK the 30 voice channel 2 Mb/s circuits have proved immensely popular, because they were tariffed at the equivalent cost of about 14 individual leased-line circuits.

Tariffing is just as sharp in the USA. In New York alone there are more than 100,000 1.5 Mb/s circuits which are shared between public (telco) and private use.

It makes good economic sense for medium and large-sized companies to replace groups of individual voice or data circuits with one or more 1.5 or 2 Mb/s digital circuits—saving money and acquiring spare capacity at the same time.

For the larger, multinational companies it has even been worthwhile to arrange permanent leases on one or more European standard 2 Mb/s or American 1.5 Mb/s circuits across the Atlantic.

> For the television broadcaster, however, these 1.5 and 2 Mb/s circuits are small change—their own requirements are a hundred times greater.

Although the generation of television programs was originally an entirely analog picture process, recent years have seen the introduction of more and more digital picture processing "black boxes" both in the TV studio and especially in the post-production areas. The more poetic broadcasters refer to these developments as the creation of digital islands in an analog sea. The size of the islands continues to grow—some islands have already joined up with their neighbors.

This has primarily come about through the commercial success of an internationally agreed digital television picture storage standard which is cryptically referred to as the CCIR Recommendation 601.

## Recommendation 601

The European 625 line and the American 525 line television standards were set many decades ago, long before digital techniques became commonplace. The early designs of the late 1930s are still considered to be quite brilliant for they exploit analog electronic techniques to the fullest.

It is customary to express the "definition" of each analog television system in terms of the visibility of closely ruled adjacent black and white lines on a test card. Across the width of the screen the American NTSC TV system can show, or "resolve," a total of about 330 vertically drawn black or white lines whereas the European PAL TV system can resolve a total of about 420 such lines. Any attempt to show any more lines or to put a few lines closer together just results in a blur.

If we are to move the television picture from the analog to the digital bakery then it is important to decide on how many ways we are going to slice the television "cake." In particular we must decide into how many small pieces or picture elements—pixels—we are to divide each horizontal line of the picture.

To a first approximation the number of horizontal pixels or sample points is very nearly the same as the number of lines of definition. A digital resolution of 330 horizontal pixels will limit us to 330 lines of definition, fewer pixels will limit us to fewer lines of analog definition on the test card. It is therefore important to choose a horizontal sample number that is set somewhat higher than we would expect to be able to transmit in practice.

CCIR Recommendation 601, or Rec 601 as it is more affectionately called, sets a high quality *digital* television studio picture standard for both the European 625 line and American 525 line TV systems.

Quite remarkably for any international agreement it has been able to define the same horizontal resolution for both the European and American broadcast systems—a total of 720 sample points or pixels to a TV line. In practice, it has proved convenient for the broadcasters to leave the last 20 or so of the 720 available points unused in order to enjoy a better technical "fit" with their existing European PAL and American NTSC TV standards. In the digital domain, which is currently restricted to the studio and the post-production area, the broadcasters are thus able to achieve a much higher standard of definition than can be transmitted to our homes by either American NTSC or European PAL.

If the horizontal resolution has been agreed at 700/720 pixels, what of the vertical resolution?

When the analog 525 and 625 line television systems were first designed it was not possible to cram every line onto the television screen. About 10 percent of the lines cannot be displayed but are "lost" elsewhere in the system. Thus, in practice, the 525 line TV system can only display about 480 "visible" lines and the 625 line system can only manage 576 visible lines. See Figure 1-3. The Rec 601 standard knows that 10 percent of the lines never see the light of day so it only stores the visible lines—480 American lines and 576 European lines. These "visible line" numbers are to crop up again and again in our discussion of television systems.

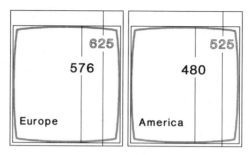

**Figure 1-3**  Visible and invisible television lines.

Rec 601 defines exactly how we must store both the level of gray, termed the luminance level, and the color, termed the chrominance value, of every pixel in every frame of the picture as an 8-bit number—2 x 8 bits = 16 bits in all.

To see these individual stored picture frames as a moving sequence we need to "play" them back at the right speed—25 pictures per second in Europe and 30 pictures per second in America. This sets the pixel rate, i.e., the number of pixels per second, that we expect to play out from the store. The maths are quite straightforward.

- In Europe there are 720 horizontal by 576 vertical visible pixels in each picture and there are 25 pictures per second. A little arithmetic sets the pixel rate at 10.368 million pixels per second and hence the data rate of the bits, which is 16 times higher, at 165.888 Mb/s.
- In the USA there are 720 by 480 visible pixels in each picture and there are 30 pictures per second. It is indeed a pleasant surprise to discover that the American pixel rate is also 10.368 million pixels per second and that the data rate is also 165.888 Mb/s—identical to the European standard.

Hence the universal appeal of Rec 601.

In practice we have to add back in not only the invisible lines at the top and bottom of the TV picture but also the invisible black margins on each side of the picture. These overheads raise the pixel rate of both the European and American TV systems from 10.368 million to an agreed speed of 13.5 million pixels per second and the corresponding data bit rate to the usually quoted figure of 216 Mb/s—though the visible picture content remains, of course, at just under 166 Mb/s.

> Thus the broadcasters, in both Europe and America, have
> found it most convenient to adopt 216 Mb/s as the universal
> data speed for moving and manipulating high quality 525 and
> 625 line digital pictures around the studio.

Broadcasters often have to mix different TV pictures together. The more sophisticated picture mixer equipment usually first separates the gray, luminance signal and the color, chrominance signal from the incoming NTSC (or PAL) signal before undertaking any complex intermixing.

Once the pictures have been artistically mixed together the resultant picture is recoded into its more familiar NTSC "composite" form. This process works well if only one or two "mixes" are required but becomes tortuous if the signals need to be remixed a number of times.

In each pass through the mixer the TV picture must first be separated into its luminance and chrominance components and then recombined. Neither the strip down nor the reassembly process can be entirely free of distortion and so the quality of the TV picture gradually deteriorates. It would clearly be much better if the picture could remain in its stripped down component form all the way through the studio rather than suffering many unnecessary recombinations and separations en route.

Broadcasters have thus welcomed Rec 601 with open arms because it always stores and transmits each picture element as two separate values: its luminance value and its chrominance value. Mixing and remixing of TV signals can now take place at component level and the intermediate results can always be stored and forwarded in their component form rather than being needlessly recoded and decoded.

But there is a small cloud on the Rec 601 horizon.

Sooner or later the mixing and remixing of the Rec 601 video signals will lead to the same "rounding error" distortions that occur in the digital sound studio and in the less expensive "composite" digital NTSC and PAL video mixers. We know that the audio engineers intend to get around the effects of this distortion by extending the sound sample length from 16 to 20 bits. We also know that TV engineers have adapted the same bit extension technique to the conventional composite NTSC TV signals and would now like to do the same to the video component format—to extend both the luminance and chrominance sample sizes from 8 to 10 bits. The television industry has yet to decide whether Rec 601 should be permanently extended to include the option of storing the TV picture as an enhanced format of 10 bits luminance and 10 bits chrominance.

> But what happens when the digital TV picture tries to leave the safe environs of the studio or the post-production editing suites?

We know that the telephone companies enjoy a near monopoly for connecting studios and transmitters together and that their "digital hierarchy" networks were set in place long before Recommendation 601 was ever conceived.

Up to now the telephone companies have offered the broadcaster specially engineered expensive conventional analog video circuits. Broadcasters complain bitterly about the cost of such circuits. There is now a second option, the high speed 140 Mb/s digital circuit, which offers both the telco and the broadcaster more flexibility. It should prove less expensive than its analog counterpart and is now considered the preferred choice for new video circuits.

However, before the broadcaster can take the telephone company's kind offer of a digital circuit he must first squeeze or compress his 216 Mb/s digital video signal to make it fit into the telco's 140 Mb/s "pipe." In order to see how this might be achieved let us return, for a moment, to the digital bakery.

### Packing the cake

We know that the digital cake comes to no harm when each piece is individually wrapped. We also know from experience that wrapping every item individually must take up a lot of room in the packing case. Is there any way of reducing the bulky digital packaging to the same level as analog packaging?

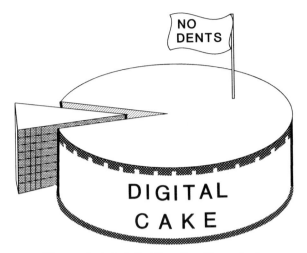

**Figure 1-4** Subdividing the slices of cake.

Fortunately there is a wide range of tricks that can be played in order to reduce the packaging needed.

> The first family of tricks is to bundle as many of the standard sized pieces of cake together wherever possible.

As we first slice the digital cake into millions of pieces ready for packing, we might notice that many of the slices have a characteristically similar appearance. Although some slices are a random jumble of cake and topping pieces, for many slices we notice that we encounter a good "run" of cake pieces before we reach a piece of topping. When we come to pack the pieces we can take advantage of these "runs" of cake pieces.

For example, we could create a standard "bulk pack" size that is made up of a "run" of 25 pieces of cake taken together rather than packing the pieces one by one. We might then send "One bulk 25-cake + One cake + One topping" in order to convey an order for "26 cake and 1 topping" more efficiently. This bundling technique is termed run-length encoding and is both powerful and very flexible.

Run-length encoding is the basic technique used in the now ubiquitous Group 3 facsimile machine which, on a good day, can send a written page around the world in 20 seconds. The earlier

Group 1 and Group 2 fax machines took many minutes a page and were never a commercial success.

We know, however, that some pages just seem to rattle through whereas other pages seem to struggle through the fax machine. It does not take long to notice that it is what is actually written or typed on the transmission page which determines how long the page will take to transmit. A nearly empty page takes no time at all whereas to fax a cutting from a newspaper takes much longer. After a lot of trial and error the run-length or bulk-packing scheme in the fax machine has been adjusted to transmit "ordinary" pages of text quite quickly whereas unusual pages will take much longer to transmit.

For example, in the digital bakery our slice of cherry cake might often naturally divide into sequences of "26 cake, 1 topping." In this case the idea of bulk packs of 25 pieces of cake at a time would make good sense. If, however, the usual cherry cake sequence was "24 cake, 1 topping," then our previously chosen standard bulk pack size of "25-cake" would no longer pay off but might even become a liability.

It is therefore very important to check the size of the likely "cake-topping" sequences before deciding that, for example, a 25 piece bulk pack size of cake is a good idea.

The digital bakery will probably be asked to make a whole range of cake sizes. If the natural packing sequences were likely to be both "24 cake, 1 topping" and "26 cake, 1 topping" then the smaller bulk pack of 24 rather than 25 pieces of cake would suit both cake sizes better.

One popular method of dealing with this variability in the length of the "cake-topping" sequences is just to send information about their differences.

The first sequence of "26 cake, 1 topping" could thus be followed by instructions such as "same again" or "please add one more cake piece next time." At first sight this differential coding technique appears to be very effective but it is prone to error.

For example, we might consider that road directions such as "Turn left, straight on, take third right at the lights" will offer our visitors a clear and concise route from the freeway exit to our house—but our visitors get lost all the same.

The inclusion of absolutes such as "Turn left into Main

Street" and "Third right at the lights into Springtree Avenue" makes all the difference to the intelligibility of the directions as any errors in navigation are thereby self-correcting. Thus differential coding can provide a useful reduction in bit rate but must be used sparingly if it is to remain effective.

It is clear that quite a lot of work has to go into "tuning" or optimizing the size of the bulk packs and it is probably not possible to satisfy everyone.

The inbuilt set of bulk-packing rules in the fax machine—more accurately termed the compression algorithm—can sometimes be totally defeated or "busted" if we happen to transmit an unusual message. For example, one of the best methods of busting the Group 3 facsimile algorithm is to try to transmit a page of ordinary school book graph paper. The closely ruled squares are quite unlike any normally encountered message and the fax machine's "Group 3" algorithm has a very difficult time with the page.

## Bulk pack television

If we apply these bulk-pack or run-length coding ideas to the TV picture it is quite easy to reduce the digital picture bit rate from the Rec 601 speed of 216 Mb/s to the telephone companies' 140 Mb/s.

- On the one hand we only need to reduce the visible picture elements which constitute about 166 Mb/s of the 216 Mb/s data bit stream.
- On the other hand we might later appreciate some spare room or headroom so as to be able to include other signals in the 140 Mb/s circuit.

We therefore arrange to squeeze the raw 216 Mb/s Rec 601 rate just that little bit more in order to achieve a picture bit rate of about 125 Mb/s. This leaves a spare 15 Mb/s which can be used for error correction, studio stereo sound, closed captioning and all the many other housekeeping signals that need to be sent to the same destination.

> Thus we have compressed the raw digital video signal from 216 Mb/s so as to fit easily in a 140 Mb/s bearer circuit.

There is virtually no degradation in the quality of the picture from that of the original Rec 601 image. In this way the signals can be transmitted around the country and around the world without any further loss in quality.

The 140 Mb/s circuits can now link the broadcasters' previously disparate digital video islands together. They are no longer bound by the familiar shortcomings of the USA NTSC or European PAL TV transmission formats but can take advantage of the far more useful Rec 601 component video format.

The success of the run-length encoding techniques at 140 Mb/s naturally leads us to ask how much further we might push the idea of squeezing more TV pictures down the telephone companies' digital circuits.

Can we, in the words of the popular game show, double our money?

It has been found that it is relatively easy to double the compression performance in order to squeeze two digital video signals into the 140 Mb/s bearer. There may be a slight loss of quality on some "difficult pictures" but this is offset by the considerable reduction in cost by transmitting not one but two quite separate TV pictures in the 140 Mb/s circuit.

The transmission of these "contribution quality" TV pictures at about 70 Mb/s does, however, come close to the performance limit for this relatively simple compression technique. It may prove possible to extend this just a little more if Rec 601 were to be permanently upgraded from 8 to 10 bits per pixel. However, if we want to reach the next milestone in the telephone companies' digital hierarchy—45 or 34 Mb/s—then we need to think again.

### Home cooking

We know that the digital cake can be safely shipped if the individual pieces of cake are either wrapped separately or are bulk-packed. Although a particular bulk packing method might suit one particular cake very well we would not be surprised to learn that it may fail to suit another. Clearly the wider the range of cakes produced by the bakery, the more difficult it is to optimize the bulk packing sizes. There is thus a limit to how effective our bulk packing methods can become.

Is there another way of solving this problem?

Yes, the second family of tricks is to send the recipe and the ingredients rather than the pieces of the finished cake itself.

"Take 4 eggs and 2 pounds of flour" seems much quicker and easier to specify than trying to describe the made-up batter in the cake. However, this short cut can only work if we know for certain that the batter was originally made that way and that we are sufficiently competent to follow the recipe.

This sort of recipe making is quite straightforward and, in the hands of a good cook, can lead to excellent results.

The hard part is to undertake the inverse process.

If we are presented with a made-up batter are we able to say for certain whether the batter used up three eggs or four?

Perhaps the only way to be really sure is to try each permutation of eggs and flour in turn until our new batter matches the original. Although it might take a great deal of trial and error to determine the exact recipe, once we know the ingredient quantities it is then easy to make more batter just like, or even better than, the original.

It is also now practical to send more than one recipe:

- one cake recipe for the plain cook, another for the cordon bleu cook.
- one audio recipe for the ghetto blaster and another for the better quality domestic Hi-Fi installation in the den.
- one video recipe for today's color TV set and another for tomorrow's high definition television receiver.

But we are jumping ahead. We must first find a relatively easy method of reverse-guessing the recipe for the batter.

## Joseph Fourier

Nearly 200 years ago the brilliant French mathematician and Egyptologist, Joseph Fourier, set the guidelines for what is now both revered and feared by aspiring mathematicians—the Fourier Transform. In mathematical terms he demonstrated how events that occurred in "time" could be expressed in the "frequency" domain and vice versa.

He showed that it was possible to perform mathematically what music students have always had to accomplish the hard way by "training" their ear.

We know that budding musicians take a series of ever harder examinations to test their competence at their chosen instrument. Each examination includes an aural test in which the student must identify an unknown sequence of two, three or more notes—a chord or an arpeggio—which is played to the student by the examiner.

The chord may be in a major or minor key, diminished or augmented, with perfect fifths or major thirds. As some of us will remember from personal experience, it takes a lot of practice to get these right. Some musicians are also blessed with "perfect pitch" and can tell, without reference to a tuning fork, whether each note was an A or a G, a Cb (C-flat) or a B.

> This might be hard on our ears but is child's play for the Fourier Transform.

The Fourier technique can "listen" to a piece of music and can then literally "transform" the sounds back into the musical score from which they were played. If the recording is of piano music the Fourier Transform can specify the correct placings and lengths of every incision to be made on a paper pianola roll which could then be used later to duplicate the music on a player-piano. Apart from the good citizens of Melbourne, Australia, the major market for pianola rolls may have passed but its electronic equivalent is the clue to "compressing" the sound.

At first the Fourier technique was too cumbersome to use—an enormous number of mathematical multiplications and additions were needed to reach the final result. However, with the invention of the digital computer, the method proved of immense interest to computer scientists who could now leave all the hard number crunching to their machines.

### The Fast Fourier Transform

The Fourier Transform remained of academic interest until about 25 years ago when researchers discovered some very useful ways of performing major short cuts in the calculations. In simple language they found out how to cheat.

A 100-point sequence no longer needed to take 10,000 calculations but could be achieved within a few hundred. More important, a million point sequence, which might represent 30 seconds of music, no longer needed the unachievable target of 1,000 billion calculations but could be done in only a few million.

Computer chips and this shortcut method, dubbed the Fast Fourier Transform or FFT, suddenly made it possible to perform these calculations in what computer buffs kindly call "real time." In other words it was now possible to print the note on the musical score within a few tenths of a second of that note first sounding.

All of a sudden the FFT allowed us to switch easily from "time" to "frequency"—from seeing only a meaningless repetitive squiggle drawn in time across the oscilloscope screen to being able to positively identify the squiggle as real musical notes—the frequencies that constitute, for example, the first inversion of the major chord of D. This is illustrated in Figure 1-5.

Although this is all good news for music students it is only half the story.

Imagine what happens if we perform the calculation twice—by doing a second FFT calculation on the results of the first FFT calculation.

We will witness two transforms—from the music to the musical score and then back to the music again.

We will, of course, be back where we started.

Or not.

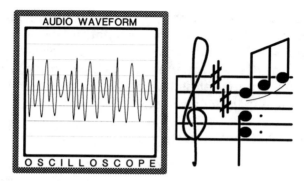

**Figure 1-5** The Fourier Transform can turn the oscilloscope squiggles into musical notation.

We too could cheat.

After the first FFT stage we could "edit" the musical score. We could decide to change some of the notes or the players on the score. We could shoot the double bass player and substitute a bassoon player in his place. We might belatedly decide, for example, that Glenn Miller got it all wrong and replace his clarinet with a tenor saxophone.

Why stop there?

We could change the key.

Was it not the pop singer Shirley Bassey who shocked the world of classical music by suggesting that some famous operatic arias would suit her own voice very well if they were transposed down a few notes.

Shock, horror, but very easy to do!

We could even take over the baton of the conductor and change the dynamics and the speed. We could massage the score to make the song last an extra few seconds in order to fit the TV schedule. Easy.

All these things are now done every day—for it is no more than the musical equivalent of word processing.

### Combing the sound

Just as useful is the ability to take away all the un-musical sounds—those sounds that have hidden themselves in the gaps between the notes on the music stave—those gaps that are normally only visited by an out-of-tune piano or a poorly trained singer who strays off note.

We are most aware of audio noise—hiss—when it falls in the gaps not used by the music. Just like the initial blue wash of an artist's watercolor sky the background hiss is everywhere—it is just that we are no longer able to perceive it when the wanted musical notes are placed in the foreground.

The hiss on an old recording will not be missed—we can chop it out of the score and it is gone. The mains hum and the crackle which also show up on the score as unmusical "glitches" can go too.

We can comb and recomb through the score and get rid of more and more of the extraneous noises that might sit in the gaps between the notes. When we replay the music it is squeaky clean. The background hiss has gone.

> But have we overdone the combing and made the music sound too clinical, perhaps too surreal?

"No problem!" we might exclaim. Surely modern technology will permit us to add in a bit of artificial, well-behaved noise—the blue rinse of the watercolor painting—which can blur the sharp edges of the sound and hide the joins.

Or have we come face to face with a real problem?

### Join the choir

We know that an amateur choir can be significantly improved by the welcome recruitment of a few strong voices. The new singers seem to impart an added depth to the sound which is almost out of proportion to their numbers. The choirmaster will, no doubt, be delighted with the improvement but would not dare suggest that the weaker voices are now redundant and should retire gracefully. He would not agree that the strong voices had masked the weaker ones—he would probably be more diplomatic and say that they have enhanced the contribution made by their weaker brethren.

Yet a check of the microphone signal will clearly identify the strong voices and give far less importance to the weaker ones.

Will the FFT combing process make the same mistake?

Will the FFT inadvertently dismiss the weaker voices in the same way as it would comb out a slightly off-key noise and eliminate them from the score? If the combed and processed sound should then appear a little surreal, will some additional added-in noise—the artist's blue rinse—or the introduction of a small amount of echo help to restore the realism?

The acid test is this. Will the trained ear notice the difference between a strong-weak amateur choir and a strong-nothing choir that has been "improved" with the added "blue rinse" noise? Even more critical—will the trained ear still be able to distinguish between two nearly identical amateur choirs when they have both been subject to the blue rinse treatment?

- The answer is clearly more important to the trained classical musician than for the casual listener to pop music.
- The answer is also important to the transmission engineer because the strong-nothing choir can be digitally compressed much further than when the weaker voices must be left in.

The proponents of the blue rinse technique are reluctant to demonstrate the effect on classical music. Instead they prefer to demonstrate the "before and after" results on "easy-listening" music which was recorded with some added echo.

My own ear—which is hardly off the bottom of the learning curve—tells me that when there is some "echo" behind the singing then it is difficult to tell the difference between the two techniques.

On the other hand, when the original recording has relatively little echo the difference is easier to spot—the blue rinse gives a "fuller" sound. If we might refer back to our digital bakery for a moment, it is as though the strong voice-blue rinse technique always provides topping even though we would occasionally prefer just plain cake.

Perhaps one of the most critical pieces of test material for any compression system is music played on an unaccompanied clarinet. I have heard such material played repeatedly through a number of different proprietary audio compression systems and they all sound fine—until compared to the original recording. They also all sound different, just as though the clarinet player had swapped reeds or mouthpieces or even instruments between performances. Pity the clarinetist whose $3,000 instrument has been downgraded into a student's model and whose carefully crafted dynamics have been flattened unexpectedly. Pity too the percussionist in the pop music band whose riffs can suffer the same fate. No wonder that many drum solo recordings sound so boring when compared to a live performance.

**The TV jigsaw puzzle**

Although a sequence of a million sample points can provide 30 seconds of music this amount of data is a mere drop in the ocean for a TV signal. As we learned earlier, Recommendation 601 has an insatiable appetite for digital samples and can both swallow and deliver 13.5 million of them per second. This is an enormous number by any standards and it is not surprising therefore that FFT computer chips are only just able to tackle these high speeds.

Yet, without these chips, digital television would not exist.

No one, not even a computer chip, can take on 13.5 million samples in one go. This number must first be broken down into smaller pieces of more manageable size.

How can this be done?

Unlike the continuous sound signals, which we met earlier, we are fortunate that the television signal breaks down naturally into individual "frames" or separate pictures—25 a second in Europe and 30 a second in America.

The European version of Rec 601 defines the frame as 576 visible lines by 720 horizontal dots—a total of 414,720 pixels. This figure of just under half a million pixels looks more manageable immediately. The American version, 480 lines by 720 dots, yields 345,600 pixels—a mere third of a million pixels which looks even better.

We can next chop up the picture into relatively coarse small blocks of about 30 or 36 blocks high by about 45 blocks across. Each TV picture is thus made up of about 1,500 blocks—1,620 in Europe and 1,350 in America.

The digital TV picture is thus no more complicated than a 1,500–piece jigsaw puzzle.

A little arithmetic shows that each piece of the jigsaw is a very convenient size—16 points wide by 16 lines high, a total of 256 pixels. This particular size suits the FFT algorithm very nicely (as it can be divided accurately by a factor of two many times over).

We can easily express the digital TV transmission requirement as:

Europe—1,620 jigsaw pieces 25 times a second

America—1,350 jigsaw pieces 30 times a second

As we have come to expect of Rec 601, both TV systems thus require the same transmission rate of 40,500 jigsaw pieces per second. This is surprisingly close, though entirely by coincidence, to the domestic compact disc (CD) sampling rate of 44,100 audio samples per second.

More of this later.

Each brightly colored piece of the jigsaw contains 256 points, each of which is a 16-bit sample (8 bits of luminance and 8 bits of chrominance). The total number of bits per jigsaw piece is 4,096. We are still on track: 40,500 jigsaw pieces per second times 4,096 bits in each jigsaw piece takes us back to the visible line Rec 601 digital transmission speed of 166 Mb/s which we first derived on page 38.

Can we fulfill our promise of a few pages ago and compress this 166 Mb/s into the 45 or 34 Mb/s digital bearer circuit?

Yes. It can be done but we must first take care not to allow the TV picture to hog the whole of the digital circuit—some room must be left for stereo sound and all the other ancillary broadcast information. Let us therefore allocate the European video a net rate of about 28 Mb/s from the overall 34 Mb/s transmission rate.

Resorting to the pocket calculator once again, we can see that the 40,500 jigsaw pieces per second can each be allocated about 700 bits per piece as against the full speed un-compressed number of 4,000 or so bits per jigsaw piece.

For the European 34 Mb/s standard this is a compression ratio of about six times. For the American 45 Mb/s standard this is a little less demanding compression ratio of about five times. Both can be obtainable relatively easily with the FFT.

A compression ratio of five or six times sounds like good news. Can we try and double our money once again and push for an even higher compression ratio?

Running similar calculations for the Group 3 facsimile machines shows that we can obtain compressions of typically 20 times—even by using the regular run-length encoding techniques rather than the deluxe Fast Fourier Transform.

Why not insist on the same compression performance for the TV signal?

We can indeed push for better compression performance but there is an unexpected price to pay.

The price is that of time.

We know, from experience, that a white piece of paper will rattle through the fax machine very fast whereas a newspaper clipping takes a much longer time to transmit. Nobody minds this delay at all—what does an extra few seconds matter?

The TV engineer cannot enjoy this luxury—he has no "time" to spare. Instead he is expected to guarantee that every picture reaches the receiver right on cue—25 or 30 complete pictures a second—no more—no less. His hands are tied—or are they?

Might he have some flexibility after all?

It is here that we start to notice the differences between the requirements for the contribution and distribution qualities of TV pictures.

Each frame of a contribution quality TV picture is expected to be able to stand on its own—with no reference to or dependence on its neighbors. If, for example, the TV program is a sports event then it is important that, in the slow motion replay, each frame of the picture is as clear as can be.

This can only be achieved if the contribution link is able to transmit each individual frame of the TV picture quite separately from its neighbors. Broadcasters place a high value on the frame-by-frame picture independence of both the European PAL and especially the American NTSC TV systems and would be unwilling to lose this ease of editing.

> Thus the six-fold compression of a Rec 601 TV "contribution quality" TV picture into a 34 Mb/s format probably represents the current frontier of technology for contribution quality TV links.

However, as briefly hinted a moment ago, there remains a wide range of tricks that can be played on the distribution side of the TV studio—where we have been assured that no further picture processing will need to take place.

### Half a second

We know that the TV engineer is committed to delivering 25 or 30 TV pictures per second precisely on time. What we have omitted to mention up to now is whether these pictures are intended for immediate delivery or whether a delay of a fraction of a second might be permitted to elapse before the picture appears on our screen.

Twenty years ago there was virtually no picture delay—the TV picture was being drawn on our TV screen at home no more than one thousandth of a second later than it left the camera in the local TV studio.

Live action indeed!

Today the TV signal is delayed a few tenths of a second in its passage through the now far more sophisticated studio and transmission equipment.

The old analog sea still offers the TV picture a fast journey—it is the hopping from one new digital island to another that now slows down the TV picture.

We know that satellite transmission will add a further delay to both the picture and the sound—for we have all suffered from international telephone calls that sometimes include an annoying "echo" three tenths of a second after we speak.

Is this delay important?

Many business users in the financial sector cannot tolerate

this extra 300 millisecond delay when telephone calls are routed over a satellite—they would far prefer a terrestrial or sub-sea route.

Their reasoning is at the root of an astonishingly simple competitive advantage:

### The one syllable rule

Via a terrestrial or sub-sea cable, the familiar Wall Street staccato instructions "Buy," and "Sell," can be initiating remote action before the same sentiments have even reached the desk of the rival trader whose only contact is via the longer, slower satellite link.

For the domestic viewer, on the other hand, does it really matter that the TV coverage of the US Masters or of the World Series has reached him a half second later than it occurred in "real time"?

Deliberately building a transmission delay into the distribution system can help the digital transmission engineer enormously—he can keep it in reserve or "up his sleeve" to enable him to deal with difficult pictures.

To achieve this we must first introduce an artificial delay to the passage of the easy pictures so that they take as long through the system as would the difficult pictures. In practice the delay might be fifteen or so frames of the picture—a delay of about half a second. The difficult pictures can then be allowed to steal some of the transmission time which was allocated to, but not needed by, the easy pictures.

This unequal sharing of resources on a digital transmission link is a well-established technique, not unlike the protocol of a meal at a Chinese restaurant.

At such a meal the dishes are usually placed centrally but within easy reach of everyone. Everyone helps themselves, a few mouthfuls at a time, until the dishes are empty. Within the overall limit of the amount of the food in the communal dishes, each guest takes exactly what he requires.

Contrast this with the meal at the fast food diner where

everyone receives a standard size portion—perhaps not enough for some hungry guests but more than sufficient for others who may leave food on their plates.

> Wherever possible, telecommunication engineers prefer to adopt the Chinese restaurant protocol of allocating the maximum amount of resources across the widest range of recipients rather than making preallocation decisions—which, in hindsight, can often prove to be incorrect.

In the final chapter we will return to this idea of allocating resources more effectively.

Let's get back to the jigsaw puzzle.

Our childhood jigsaw puzzle experience reminds us that the larger and harder jigsaw puzzles always seem to contain lots of pieces of sky or trees—all irritatingly close to but not quite identical to one another.

In frustration we may have been tempted to cheat a little and force the "wrong" piece of sky into place on the puzzle. We knew this was impossible but with the TV jigsaw picture we are now able to cheat in two ways.

## The code book

We remember that there is not much difference between one patch of sky and its neighbor in the jigsaw puzzle. Agreed, some pieces might have a little more fluffy cloud in one corner than the others but most of the area of each piece has the same identical blue color as its neighbors.

> Have we not all wished that we could snip off and use just a part of a jigsaw piece in order to "finish off" the sky?

We can easily achieve the same "snipping" effect in the TV jigsaw picture if we divide each TV jigsaw piece into smaller pieces called sub-blocks.

We can subdivide the basic block of 16 by 16 pixels into a number of smaller sub-blocks which are each, say, 4 pixels high by 8 pixels wide. Across such a small area the variation in the blueness of the sky is not significant—in effect the luminance and chrominance values of the 32 pixel sub-block (4 by 8 pixels = 32) are virtually all the same.

In one stroke we have thus reduced the amount of data about the small sub-block by a factor of 32 times.

We can go further.

It is very probable that there are some more small sub-blocks of sky with exactly the same blueness not far away from the first or reference sub-block. It is therefore worth taking a careful note of the characteristics of the reference sub-block for judicious use elsewhere in the sky. It would clearly also save time if we could invent a form of shorthand which could quickly describe the sub-block rather than repeating the full description every time we need to reuse it.

We can open a shorthand note book—more correctly termed a code book—and carefully record the characteristics of the reference block under a shorthand mnemonic such as "Sky 1." We can do this a number of times, thus adding more entries to the code book. The "fluffy cloud" mnemonic could be "Sky 2," whereas darker skies could be "Sky 3," etc.

Next we must decide how many subtle shades of blue sky we need to reproduce. Do we need to define 100 shades of blue or will 10 shades be sufficient?

If we were to define only 10 shades of blue sky will the adjacent jigsaw blocks of the TV picture start to resemble a child's attempt at a collage and thereby show the joins? On the other hand, if we delineate a hundred shades of blue, will the human eye ever notice the difference between the mnemonic codes "Sky 98" and "Sky 99."

Oophs. It seems that we cannot avoid having to make the same sort of quality decisions for the TV picture as we also have to make for the transmission of classical music.

Different TV pictures will need different shorthand mnemonics and code books. It would therefore be useful if the TV compression system could provide a built-in basic vocabulary of shorthand mnemonics to which can be added special vocabularies as required.

Naturally enough, all the entries in the code book must be

in place before the picture can be successfully decoded from it. This could take a little time—the time that we have kept hidden "up our sleeve" by delaying the picture a fraction of a second.

### Fractal geometry

Thirty years ago the mathematician Benoit Mandelbrot was hired by IBM to try and explain a phenomenon dear to the hearts of all city dwellers—why do buses rarely arrive on schedule but can often be seen traveling in bunches of three or four at a time?

More exactly, IBM wanted Mandelbrot to try and explain why burst noise in digital telephone systems—the crash of dinner plates from the kitchen table—seemed to occur in bunches. IBM's own computer experience seemed to indicate that there would be no burst noise for a long time and then two or three bursts would occur in a row and the system would crash.

It was as though the digital electronic circuits knew that the old wives' tale was true—accidents always go in threes.

Mandelbrot saw a great similarity between the familiar "three-in-a-row" accident scenario and the distribution of the stars in the firmament.

He asked:

- Why is it that stellar matter bunches together in characteristic fashion rather than being more evenly distributed in space?
- Why is it that the same effect of wide gaps and tight bunching occurs in nature from the microscopic right through to the macroscopic level?

Mandelbrot was able to design and test mathematical computer models that could produce their output as a series of pictures. Carefully adjusting the numbers that were fed into the model produced very plausible imitations of everything from stellar clusters to microscopic cellular structures.

He found the shape of the natural coastline of islands, such as the United Kingdom, to be of great interest and he wrestled with the problem of how to make an accurate measure of the length of the UK coastline.

To what degree of detail should the coastline be mapped?

From space the coastline looks quite "smooth" but from an airplane we can see more detail. From a stroll along the beach we can see the individual pebbles. Each time we zoom in a little closer to the edge of the coastline we encounter less smoothness and more detail.

Moreover, each time we zoom in a little closer we also need to upwardly revise our estimate of the length of the coastline in order to take in the ever smaller perturbations in the line of the coast. This begs the question. Is there even a "true" measure of the length of the coastline?

To solve this mystery Mandelbrot asked himself whether there is a relationship between the magnification factor of our view and the length of the coastline.

This is the clue to what Mandelbrot defined as a fractional dimension or fractal. We have all grown up with the idea of two dimensions—2-D—and three dimensions—3-D—but I am sure that we have never thought what 2.1-D might mean.

Might the fractional dimension be a "roughness" index which indicates the increasing complexity of the natural line as we zoom in closer?

Mandelbrot showed that the computer's mathematical model could be instructed to draw squiggly lines in, say 2.1 dimensions and then be left to run the drawing program on its own.

The resultant picture might look like a leaf from a walnut tree or the walnut tree itself. Change the initial value of both the fractional dimension and the initial squiggle which was fed into the drawing program and the resultant picture might look like the bark of an oak tree.

This might appear to be "painting by numbers" but has Mandelbrot stumbled on the way that Nature works?

- Does an oak leaf somehow "know" that it must follow a simple internal fractal geometry program to which it must refer as it grows?
- Is this why the different sized leaves on the oak tree always

look the same though they are clearly different to walnut tree leaves?

- Is this why one ethnic group looks different to another even though we all come in different sizes?

Mandelbrot showed that the shape of natural objects—stars, coastlines, trees, feathers, even the tone and texture of human skin—can all be expressed by just a few numbers which are fed into a standard computer program. To date, this technique works best for natural things—unnatural, manmade artifacts, such as straight-edged buildings, cannot be encoded so easily into a fractal form.

Perhaps one of the hardest technical tests of today's TV systems is whether they can achieve a consistent reproduction of skin tone and texture. In the late 1950s the Europeans even went to the trouble of inventing PAL and SECAM in order to overcome NTSC's difficulty of accurately reproducing skin color.

Dare we cut this Gordian knot of technical excellence and replace the TV anchor's skin with a fractal synthesis? This is not as novel as it sounds—similar substitutions are already done in the post-production stage of TV commercials in order to remove skin blemishes far more effectively than the latest beauty cream.

**Temporal dependencies**

In the UK the subject of the weather is always a good standby if the conversation should lapse. There is an apocryphal tale which suggests that, once upon a time, there was a weather prediction competition between those who use large computers and those who use more simple means to predict the weather. Over a 12-month period the most accurate prediction came, not from the weather center computer but from the simplest of all prediction techniques. In a nutshell, the technique states that, in all probability, the weather tomorrow will be the same as today.

TV pictures are the same. Barring scene or shot changes, the current frame is often very similar to those that went before. Thus, in order to update the picture, it is usually safe to transmit the differences in the picture rather than the complete—or absolute—description of each frame. This is similar to but differs in one respect from the technique of differentially encoding audio signals.

When we encode audio we need to go back only a few samples in the time dimension whereas in the TV technique we must consider the dimensions of both time and space. We must take care to match each point on the current frame with its predecessor both in the same position in space on the screen but many millions of samples back in the time dimension.

If the new picture has only changed a little from the previous frame we can still make use of the same entries from the code book and the same parameters for generating the fractal skin texture. After all, the camera angle may have changed slightly but the sky has not changed from its light-blue color.

What we are sacrificing, of course, is the stand-alone independence of each frame—instead we must now refer back a frame or more in order to establish the current frame.

But, like any differential technique, this idea can only be used sparingly, perhaps over no more than several frames. Every so often we must transmit an absolute frame. This will update all the picture information and make sure that any errors that might have crept into the differential coding can be extinguished.

If we are given the chance to analyze a few frames of the TV pictures all together it is relatively easy to tell the difference between static and moving objects. For example, it should be relatively easy to spot the automobile which is traveling down the road. We could thus transit the first frame in full and frames 2 to 5 in difference form. The static background would remain the same in all five frames. However, each difference frame (2 through 5) would need to show both the car in its new position and, most important, the part of the static background that the moving car no longer obscures.

We might extend the differential idea a little further.

From the analysis we could, once again, identify the car as the moving part of the picture and then suggest to the "intelligent" TV receiver that it should work out for itself where to place the car over the next two or three frames. The first frame would contain the whole—the absolute—picture but subsequent frames would contain instructions such as "move the car a fraction to the right," "do it again," "do it again."

This technique is probably not very accurate for relatively large movements across the screen but could be quite useful for smoothing out any small scale jerkiness in the picture that might be seen, for example, on the rolling captions or "credits" at the end of a program. The accuracy of the interpolation can be further improved if a "preview" frame is also included just after the absolute frame. In effect the intelligent TV set is told—Here is where we start (the absolute frame) and here is where we intend to be in half a second's time (the preview frame). With the help of the intermediate frames the TV set can work out precisely how to get from A to B without overshooting the mark.

There are two other important constraints on the transmission ratio of absolute and differential frames.

### Changing channel

When we select a new channel on today's TV set we naturally expect that the new picture will appear immediately. Tomorrow's TV set may be different. If the digital television transmission format is using a mix of absolute and differential encoding then the TV set will be unable to make proper sense of the new signal until an absolute frame is received.

How long are we prepared to wait for the new picture to appear—a quarter, a half or a full second? For example, if we transmit an absolute frame every 12 or 15 frames—every half a second or so, the average waiting time will be about a quarter of a second. Will this prove sufficiently fast for the so called couch potato—the reclining viewer who changes from one TV channel to another every few seconds?

### Picture search

In the early days of video cassette recorders there was no picture search facility—instead we had to search blindly for the start of the recorded program by the judicious pressing of the "rewind" and "fast forward" buttons. The situation improved rapidly with the introduction of "picture search"—the ability to speed up the normal replay speed by about five to ten times. The high speed picture quality is not good—but is sufficient to locate the start point of the program. During this operation we see parts of a number of adjacent TV frames—the bottom portion of the high speed picture will be taken from a frame some five to ten frames ahead of the top part of the picture.

If we take care, we can provide the same high speed picture search effect in the digital TV transmission signal. This can be achieved by providing each part of each TV frame with some low definition "absolute" coding clues that can still be decoded when the tape is run at a higher than normal speed.

### Dropping the plates

Burst noise, like the dropping of the pile of plates from the kitchen table is an accidental occurrence. Although we can do our best to avoid accidents we know that they cannot be avoided in their entirety.

This is why we insure our belongings, automobile and home, each for the payment of an annual premium. Part of the bargain with the insurance company is that we should first take reasonable precautions with our belongings—perhaps by fitting stronger locks to the doors of our home or regularly checking the tires and brakes of our car.

If we cannot avoid the occasional crash of plates from the kitchen table, we can at least try and mitigate the damage when it does happen. One way we could do this is to stack and mix the plates in a different order. Instead of putting all the dinner plates in a single pile we might instead mix just a few of the dinner plates with a few side plates and saucers.

To switch metaphors—we are no longer putting all our eggs in one basket. When the crash finally comes the pile of stoneware will still get smashed but the effect of the damage will be limited.

We can apply the same "stoneware shuffling" technique to the digital television signal by placing some of the information about one part of the TV picture in different frames or different parts of the same frame.

In practice this technique is not as successful as it might first appear. This is because:

1. The burst noise is going to wipe out a part of the TV signal—our shuffling technique merely allows us to decide whether we lose 12 dinner plates or six cups and saucers.
2. The "three-in-a–row" old wives' tale is probably true. Murphy's law rightly suggests that the second burst of noise will probably floor us by occurring just as we are recovering from the first burst.
3. We must avoid the temptation to be too clever. To return to the trailer analogy—there is nothing to be gained by ingeniously packing into a 14-foot trailer what can go easily into an 18-foot trailer.

Our best strategy is therefore to accept the occasional burst of noise as inevitable—to take it as a blow on the chin and then to arrange to recover from it as quickly as possible.

## WHERE ANGELS ONCE FEARED TO TREAD

Ten years ago the concept of the video-phone was once again resurrected in the form of business video conferencing. At the Fourth World Telecommunications Forum held in Geneva in 1983 a number of manufacturers demonstrated that it was possible to transmit "ordinary" TV pictures through digital circuits—not at 216 Mb/s but at 2 Mb/s.

The broadcasters threw up their hands in horror.

"Appalling quality pictures"

Indeed they were, for the definition was poor and the motion was blurred.

Four years later (Geneva 1987) the same manufacturers were back with the improved models.

Picture quality?

"Just as appalling."

But there was a caveat.

The pictures were no longer transmitted at 2 Mb/s—the speed had been reduced five-fold to 0.4 Mb/s. In terms of telephone channels the TV picture now took the space normally allocated to six rather than 30 telephone calls.

Video conferencing had passed "GO."

Remember the 13,000 private 2 Mb/s circuits that have been leased in the UK and the extensive private use of the 100,000 1.5 Mb/s circuits in New York. Few of them are fully used, they were often installed because they were an excellent half-price bargain from the telephone company. Many of them have six or more spare channels that might be used for video conferencing.

> Very often the marginal revenue cost of adding a video con-ferencing facility between many business premises is zero—it only requires the addition of capital equipment to the existing high speed telecoms circuit.

But what about the appalling picture quality?

Is it really that bad?

In practice the quality is no better than we get when shooting a "vacation video" with an ordinary inexpensive, VHS camera. The quality is neither brilliant nor irredeemably bad.

I was intrigued to know what the resultant picture quality would be if the five-fold coding improvements were used to im-prove the picture rather than reducing the speed of the data link. Seven years ago this question could not be answered as the computer chips were already running flat out in providing the "vacation video" quality.

Since then there have been two further developments:

First       The regular speed computer chips can now support even more efficient compression algorithms. For the same "vacation video" picture quality, data transmis-sion speeds have fallen further, from 0.4 to 0.1 Mb/s. Person-to-person video conferencing can now be achieved over two telephone circuits rather than six. However, if the business meeting is to include a group of four or more people at each end then the 0.4 Mb/s data speed gives noticeably better quality. The tele-phone companies know that businessmen will not settle for domestic quality smudgy pictures and have majored on the provision of dial up 0.4 Mb/s service at attractive tariffs.

Second      New, purpose-built, computer chips can now perform the FFT algorithm at ever higher speeds. At the April 1991 National Association of Broadcasters (NAB) Convention in Las Vegas a number of manufacturers demonstrated real time NTSC quality pictures which were coded at between 4 and 7 Mb/s. Two years later at NAB 93 full Rec 601 quality pictures were demonstrated at 8 Mb/s. A year later at NAB 94 the rate was down to 6 Mb/s.

### Convergence

Five years ago there was a wide gulf between the broadcaster and the video conference manufacturer.

- The broadcaster set his studio standard at 216 Mb/s but would accept a data reduction to 70 Mb/s.
- The video conference manufacturer set his standard at 384 kb/s (just under 0.4 Mb/s) but could see a much bigger market opportunity at 128 kb/s (0.1 Mb/s), if and when the much hyped 128 kb/s ISDN dial-up service became generally available.

The gulf still remains—but is now much smaller.

The broadcasters now accept that contribution links can operate at 34 or 45 Mb/s and that many TV pictures can be distributed at much less than 6 Mb/s. However, some "difficult" pictures will need much higher speeds. Recent demonstrations show that a 6–8 Mb/s transmission rate should be able to provide "Rec 601" distribution quality—a definition of 720 by 480 pixels— even for quite difficult pictures such as horse racing.

Although the 384 kb/s business video conference market has flourished there has been less interest in the domestic service at 128 kb/s. This is mostly due to the difficulty in extending a telco ISDN digital service over old-fashioned analog cable into residential districts that are situated further away from the telephone exchange than the business park.

The video conference manufactures have therefore turned their attention to higher quality pictures at higher speeds. They have not lost sight of their roots—for their early products were designed to conform with the US telephone companies' original primary data speed of 1.5 Mb/s.

> The video conference manufacturer has set his sights on the CD—the compact disc.

## The compact video disc

The 5-inch domestic quality audio CD uses no data compression; the 16 bit stereo sound samples are set down at a rate of just over 44,000 per second.

A little arithmetic shows that the bit rate is:

$$44,100 \text{ samples/second} \times 16 \text{ bits per sample}$$

$$\times 2 \text{ channels} = 1.41 \text{ Mb/s}$$

What a convenient fit into a 1.5 Mb/s American digital telco channel!

Even the format of the disc is convenient. Its "container" of 44,100 samples/second can hold both the 40,500 TV picture jigsaw pieces/second (derived on page 52) and a compressed stereo sound track that takes up the remaining capacity.

The video conference manufacturer can see a big market for NTSC quality 1.5 Mb/s video. He has a ready-made medium to hand in the standard 5-inch CD which can play for over an hour. If the domestic CD player has a digital output socket, as many already do, then a "black box" decoder will provide TV pictures and sound straight into the domestic TV set.

The possibilities have not been lost on the CD maker. He knows that a 5-inch video CD is much easier to sell and store than a bulky VHS cassette. The video CD can be stamped out as fast as the audio CDs and 12-inch vinyl albums that went before. On the other hand each VHS cassette takes many minutes to transcribe from a master tape in a high speed duplicator.

The only drawback to a major new product success is the apparent limitation of 70–75 minutes playing time. Hollywood movies run for 120–130 minutes so how can the CD be made to play for this time?.

The answer is the experimental double density CD. With a little care in the mastering process the subminiature dots that form the tracks on the CD can be laid closer together. It has been discovered that most CD players can play the experimental double density discs without difficulty thereby raising the playing time from 75 to 135 minutes.

How acceptable is the picture quality at 1.5 Mb/s?

Better than one would expect.

The recording process offers an intrinsic advantage over live TV. A live TV program must be coded and transmitted as it happens. There is little time or opportunity in the live broadcast to choose the best compression technique for each part of the picture, so the nearest one may have to do. Experience shows, however, that the technical quality of a recorded TV program can be enhanced if a little more care is taken in the encoding process.

We are not surprised, for example, to hear a few bumps and coughs at the broadcast of a live performance but we would not expect them to occur in a proper studio recording. With the advantage of hindsight it has proved possible to re-compress difficult scenes so as to raise the quality of pre-recorded 1.5 Mb/s CD videos significantly above that of a "live" digital transmission at the same speed. The resultant quality of a carefully encoded movie can be better than VHS cassette.

### The search for quality

The American black and white 525 line television system was first demonstrated by Zvorykin in 1939. In the mid-1950s color was successfully introduced by means of the NTSC system that swapped color for picture sharpness. In retrospect that judgment has proved correct, everyone prefers a less sharp color picture (with 330 lines resolution) to a sharper black and white picture with 450 lines resolution.

Can we strike any other bargains with technology? For example would viewers want to swap a greater number of fuzzy pictures for a smaller number of sharper ones?

Ten years ago Europe embarked on such an experiment.

### Regular MAC

In 1982 another color TV system was born: Multiplexed Analog Component or MAC TV for short.

It was designed:

- to convey the full 700/720 pixel Rec 601 resolution right from the TV studio to the domestic television set.
- to be capable of supporting further, evolutionary improvements in both picture and sound quality.

A strong MAC signal can deliver the full 700 pixel definition whereas NTSC can deliver less than half that number of pixels. There is no doubt that the Regular MAC pictures received at home look as good as those that leave the studio. Unfortunately there is a downside, for the MAC system offers no economies of scale. The downside is that MAC takes the equivalent of 720 units of bandwidth whereas NTSC only takes 350. Are we prepared to swap two NTSC channels for one high-quality MAC channel?

When MAC was first conceived this was not thought a problem. MAC was never intended for, and does not really suit, terrestrial broadcasting where frequency spectrum (the number of available TV channels) is at a premium. It was originally intended for high-power direct to home satellite TV broadcasting where lots of wider channels can be found more easily. On satellite it can work very well indeed—it is only when it is brought back down to earth and routed the last few miles through cable TV systems that some problems arise. Its need for extra "bandwidth" is briefly explained in the technical note on the following page.

Despite its insatiable greed for bandwidth European MAC TV will not lie down and die. As we shall discover in the following pages there have been some technical developments that have made it quite attractive for the medium term. The marketplace has yet to decide whether these improvements are sufficient to prolong MAC's life or whether the arrival of digital television will soon kill it off.

Two fundamental issues are still to be addressed by broadcasters and consumers alike:

1. If our TV systems are to be improved, how much improvement and what sort of improvement makes such a change worthwhile? Do we want quality or quantity?

2. If our TV systems are to be improved, who will be the winners and who will be the losers? Can we stall objections by discovering a win–win scenario that will satisfy everyone?

The two subsequent chapters set out, in turn, to shed some light on each question. But first we need to transmit the TV picture through the air.

### Technical note on television channel bandwidth

We are quite used to tuning in our FM radio to our favorite stations which may be found, for example at 88.3 or 105.7 MHz (MegaHertz) on the dial. A quick glance at our radio confirms that the FM Radio band stretches from 87.5 to 108 MHz—a span of just over 20 MHz—and allows us to receive dozens of radio stations.

Television signals use up considerably more frequency span or bandwidth than is needed by radio stations. American television stations are allocated a 6 MHz bandwidth in which to transmit 525 line color television whereas the Europeans use 8 MHz of bandwidth to transmit the sharper 625 line PAL TV signals. Thus, if we were to shut down all the radio stations in the FM band and turn all the frequencies over to television use, we could only properly fit three American or two European TV stations in the band 88–108 MHz.

The bandwidth requirements for MAC TV are less clear cut. Theory suggests that the MAC signal should just be able to fit in a 12 MHz bandwidth (two NTSC channels) though many engineers have suggested that the allocation of a 16 or 18 MHz bandwidth (the equivalent of three NTSC channels) would ease the passage of the MAC signal through cable TV networks.

In some parts of Europe the cable systems are under-used and there is a large amount of spare frequency space which could possibly take MAC. In Germany, for example, the presently unused band 300–450 MHz, the so called cable hyperband, has been assigned for the transmission of a contiguous group of 12 MAC

TV programs. Even this plan is now in tatters. The take up of MAC has been very slow and two hyperband channels have been "temporarily" reassigned to carry more popular commercial PAL broadcasts.

In the future there are unlikely to be many vacant tracts of bandwidth just waiting to be used. It is now too large a task to start from scratch and reengineer the radio spectrum so we must face the uncomfortable reality that any improvements in television performance must be achieved within the existing standard 6 or 8 MHz bandwidth allocations.

## TRANSMITTING THE PICTURE

Digital television transmission is not new.

Twenty years ago the British public were first introduced to teletext, a new data broadcast medium which provided many hundreds of pages of news and information. Just like closed captioning these signals would lie forever hidden behind the normal TV picture unless we bought a special teletext TV set to decode and see them.

In the UK teletext has been a great success. Forty percent of homes have chosen to buy TV sets with teletext, even though these sets cost about fifty US dollars more than the regular TV model.

The main use of teletext is for closed captioning—subtitling—for the hard of hearing though there are many other services that have proved popular. Breaking news on teletext is just that for it is taken direct from the wires of the press agencies. Foreign exchange rates are updated every 2 to 3 minutes as are the Futures market prices. Teletext also offers travelers real time airline arrival and departure times thanks to the information taken directly from airport computers around the UK.

The teletext signals are hidden in the black margin of invisible lines that lies just above the top of the visible TV picture. There is not enough room in this hiding space to put all the pages of information that we might need immediately so we often have to wait a few seconds while our chosen page comes around—just like waiting for a favorite spot on the carousel ride at a fairground.

As the size of the teletext hiding space at the top of each picture is no more than a thousandth of a second long it is important to cram in as much information as possible into this short time span. Just like the rapid burst of fire from a machine gun, the digital transmission burst speed was set as high as could safely be dared in the mid-1970s when the system was first deployed. The teletext data burst speed is thus an astonishingly high 7 Mb/s. However, each burst only lasts for a thousandth of a second every time a new picture is transmitted, that is, 25 times a second.

For the past 20 years this 7 Mb/s signal has reached over 99 percent of the UK population.

UK viewers know from personal experience that the UK teletext service is quite reliable. Given a good picture from a good antenna then the teletext pages are likely to be received without errors, those irritating square blobs that occasionally appear at random in the picture. In the fringe area, however, blobs do crop up frequently and quickly render the teletext page unreadable.

With hindsight it is now apparent that European teletext has one major drawback for its standards were set long before video cassette recorders became popular. Because of its high data speed teletext cannot be recorded successfully on a VHS video recorder—though the higher quality Super VHS (S-VHS) format is just able to capture the digital signals. In order to overcome this difficulty the newer American closed captioning system uses a much lower data rate which is compatible with VHS recording. By dropping the data speed in this way the subtitled recorded tapes can be enjoyed time after time by the hard of hearing. However, the lower data rate precludes the provision of many other real time data services which Europeans have come to value and enjoy.

The hard truth is that teletext either works or it does not—but this is just what we would expect of any digital system.

Once teletext became established, broadcast engineers asked themselves how might they build on their success. There were a number of possible enhancements.

The most obvious improvement was to expand the teletext invisible line hiding space by lengthening the time window, even at the expense of encroaching into the visible picture area.

Perhaps the ultimate application of this idea is seen when the teletext data space is allowed to expand so as to displace the usual visible TV picture entirely. Such an idea has been implemented by the Singapore telephone company where "full screen" data transmission provides both a public teletext and a private data terrestrial broadcasting service. This data service emanates from a TV transmitter which broadcasts no meaningful television pictures or programs at all, only a continuous teletext-like 7 Mb/s data stream.

If the complete loss of all the TV picture might seem too high a price to pay, what of the host of intermediate solutions in which we are able to choose to lose only a little of the top and bottom of the visible picture?

## Wide-screen television—cropping and topping

In its search for new products the TV industry has come to believe that a more rectangular television display is more appealing than our old-fashioned TV screens which have always exhibited an aspect ratio of 4 units wide by 3 units high, or "4 by 3" for short.

After much experiment it has been decided that the "new look" shape of the wider TV screen should be twice as rectangular as before, i.e., "4 by 3" times "4 by 3." A moment's mental arithmetic provides today's magic screen number, the "16 by 9" shape of the wide screen.

Although a number of manufacturers have invested in and started to produce this new shape display tube for a new range of television sets, the vast majority of ordinary TV sets will continue to remain in use for many years and will retain the conventional "4 by 3" display shape.

It is therefore rather important for us to decide how to arrange the co-existence of the 4 by 3 and 16 by 9 pictures.

There appear to be four principal ways in which this can be

done. Unfortunately, none of them has yet proved entirely satisfactory.

### *European PALplus—the super letter box*

The first method is to transmit the TV pictures in what is
generally described as the "letter-box" format. This is the oblong
picture shape which is used when a wide-screen feature film is
broadcast and in which we see black bands at the top and bottom
of the picture.

From what we have just learned, these black bands can be
put immediately to good use. We can commandeer these otherwise
abandoned and unused black parts of the picture to transmit
teletext-like data signals. These signals can be used to enhance
the picture so as to fully refill the screen of our new, expensive
wide-screen 16 by 9 television set.

The intrinsic disadvantage of this approach is that the existing vast numbers of ordinary 4 by 3 television sets do not get a
full picture—they get the oblong letter-box shape instead.

> The letter-box scheme is relatively simple and could be used
> to enhance both our existing PAL and NTSC TV systems.

The main drawback of this method is that many people with
conventional 4 by 3 television sets may not like to watch the
letter-box format. We would naturally prefer to watch a picture
that fully fills the screen and, at present, the broadcaster does his
best to provide us with such a picture. Thus our favorite televised
feature film may start in letter-box format but will slip into a
full-screen format after the title sequence has finished. It is only
at the end of the feature film, when the hero has saved the world
and the credits roll, that we are likely to notice, with a start, that
the film has jumped back, once more, into the letter-box format.

The jury is still out on whether we might all get used to and
accept a letter-box format on all our 4 by 3 TV programs, just in
order to satisfy the initial few thousand viewers with wide-screen
16 by 9 TV sets. We must therefore ask:

> Is it possible to achieve a measure of normal (4 by 3) and
> wide-screen (16 by 9) compatibility by some means that does
> not require us to "letter box" all our favorite TV programs?

Yes—we can borrow the same technique of zooming-in that is used every day for filling our normal 4 by 3 screens with the main action from the wide-screen feature film. It is, of course, quite easy to make the conventional 4 by 3 picture look fine on the 4 by 3 TV set for it is done every day. We will, however, now need to find another way of conveying the extra width or "wings" or "side-panels" of the wide-screen 16 by 9 picture onto the 16 by 9 television receiver.

### Two American enhancements

In 1987, before digital television hit the headlines, the American proponents of new advanced television broadcasting systems suggested a number of ways in which a 16 by 9 picture can be provided alongside the 4 by 3 picture. They outlined two main avenues of approach:

First      They described various experimental ways of squeez-
           ing the additional "wing" and enhancement informa-
           tion into the existing NTSC TV channel so as not to
           upset the normal 4 by 3 picture.

Second     If this re-packing technique did not work, they sug-
           gested the provision of a second, low-power, **augmen-
           tation channel** that would carry the enhancement
           information alongside the existing high-power TV
           channel. As we shall soon discover, it was this search
           for augmentation channels in the crowded television
           frequency bands which led the US television industry
           to its current interest in digital television.

### Wide-screen MAC

The fourth approach to wide-screen compatibility is offered by the European MAC TV system. If regular MAC is to be used for wide-screen pictures then the 700 horizontal pixels are simply assigned to the full width of the wide-screen 16 by 9 picture. On a new 16 by 9 television set the picture will look fine.

When driving a conventional 4 by 3 TV display, however, the MAC set top decoder is simply told to "lose" the edges, the wings or side panels, of the wider picture. This is a most elegant solution for it is simple to implement in hardware and kills two birds with one stone. For the purist, however, it has a slight disadvantage when compared to conventional 4 by 3 regular MAC broadcasts.

The disadvantage is as follows. In regular wide-screen MAC only three-quarters of the 700 pixels will be assigned to the conventional 4 by 3 picture: the other quarter measure must be assigned to the left and right hand "wings" that make up the wider picture. Thus only 525 pixels are assigned to the central 4 by 3 picture and so the horizontal resolution is only 50 percent better than a standard 330 pixel 4 by 3 NTSC picture. Will the average 4 by 3 viewer notice that his MAC screen is not a full 100 percent better than NTSC? Probably not.

For the purist MAC TV engineer there is a straightforward solution to this loss of horizontal resolution—we should just "sharpen" the picture by 30 percent before transmission. If this is done the wide-screen 16 by 9 picture will now enjoy a 960 pixel resolution and the conventional 4 by 3 picture is restored to its normal 700/720 pixel Rec 601 resolution.

By this improvement we have done three things:

1.   We have very nearly described High Definition MAC— HD-MAC.
2.   We have increased the required MAC bandwidth from 720 units to 960 units.
3.   We have invented a super-set of Rec 601 in which we must record not 720 but 960 pixels per horizontal line.

Gone is the chance to fit MAC into two NTSC channels, we now need to sacrifice the equivalent of three NTSC channels for every wide-screen MAC transmission.

### Enhancement signals

In the last few pages we have fleetingly referred to the "enhancement signals" which may be conveyed in either the teletext hidden space or by other, equally clandestine, means which must not interfere with but can improve the normal TV picture.

The enhancement signals are no more than the application of the digital compression techniques which we encountered in the previous section of this chapter. These are used to sharpen the conventional TV picture and produce improvements which are variously described as either Enhanced Definition (EDTV), Improved Definition (IDTV) or High Definition (HDTV) pictures. They differ from the previously described "straight" digitally compressed television in that, rather than starting from scratch, they build on and enhance the existing, conventional TV picture.

In the research laboratory it has been possible to demonstrate that quite modest amounts of enhancement can provide a marked improvement in picture definition. However, it has proved difficult to achieve the same results in practice, especially when reception of the basic "over the air" TV picture is marred by noise or ghosting.

Life becomes even more complicated when the enhanced or improved or high definition picture is to be shown in a 16 by 9 wide-screen format whereas the basic picture, on which the enhanced picture is to be based, is only transmitted in its conventional 4 by 3 format.

> By analogy, it is sometimes easier and quicker to knock down the outhouse and start again from scratch rather than trying to restore and extend the crumbling brickwork.

## Comparing systems

At this stage we may make our first provisional assessment of straight digital compression TV versus the enhanced analog TV techniques. For convenience we can take our picture quality reference as the 700/720 pixel horizontal resolution of CCIR Recommendation 601.

*Regular MAC TV.* This system transmits the TV picture in analog form, though, as we shall see shortly, MAC has an ancillary digital component as well. The basic MAC format provides us with Rec 601 quality whereas any digital enhancement signals that may be added to HD-MAC will further improve the resolution.

MAC TV takes the equivalent of two, possibly three full-power NTSC TV channels.

*Enhanced PAL/PALplus.* This system transmits the basic 420 pixel letter-box format PAL picture in analog form. It adds teletext-like data signals in the unused lines of the picture in order to raise or enhance the resolution to 700 horizontal pixels. At a guess, in order to effect this improvement, we need to run the teletext-like data enhancement signals at a data speed of about 1 Mb/s.

Enhanced PAL takes up only one (existing) full-power PAL TV channel, but the increased resolution on the new wide-screen TV sets is obtained at the expense of seeing a letter-box format on the vast majority of existing old-fashioned TV sets.

*Enhanced NTSC.* This system will also require an enhanced data assistance channel of about 1 Mb/s. If consumer pressure dictates that the letter-box format is unacceptable and is not to be used then there will be no room for the 1 Mb/s enhancement channel in the basic TV channel. The extra enhancement data will need to be broadcast instead on a separate, nearby TV channel.

Enhanced NTSC takes one (existing) full power and one (additional) low-power enhancement TV channel.

*Straight digital compression TV.* This system requires a data speed of about 7 Mb/s in order to provide full Rec 601 700/720 pixel quality. This could be broadcast from an existing full-power standard NTSC or PAL TV transmitter in a "Singapore style" full screen teletext format.

The choices seem straightforward:

1. Given good quality television cameras and well maintained equipment both NTSC and PAL TV systems already offer more than adequate 4 by 3 TV pictures. Do we need anything better?
2. Regular MAC TV pictures are twice as good as NTSC but need two or perhaps three channels in order to achieve this gain in quality.

3. Digital enhancement channels can further improve all three (NTSC, PAL and MAC) systems.

4. An all digital 7 Mb/s TV transmission system can match MAC quality in just one NTSC channel—but at what cost and effort? We will need some very positive reasons if we are to follow this route because the development costs could be appreciable.

## MAC's nine lives

When the United States first agreed to adopt the NTSC color standard in 1952 it set the benchmark for every color television system that followed. In the late 1950s the Germans invented PAL in order to automatically correct the change in hue that can occur between different NTSC transmissions. The French invented SECAM to overcome the same problem. As a result European TV sets have no need of (and are therefore not fitted with) a Hue control. It is a credit to their design that, despite a few shortcomings, all three systems have remained in the marketplace for thirty or more years without change.

MAC has not enjoyed such stability. In its ten-year deployment it has spawned many variants.

To date we have encountered Regular MAC, Wide-screen MAC and HD-MAC. Yet we may read in the popular press of even more varieties of MAC, such as D-MAC and D2-MAC. The technical press might even include an occasional reference to B-MAC and C-MAC.

Which MAC is best?

- **B-MAC** is the oldest form of MAC in the marketplace. It is now produced by Scientific Atlanta Inc. and has enjoyed worldwide success in specialist markets. It offers better picture definition than NTSC or PAL.
- **C-MAC** was the first MAC to be based on the 700/720 pixel resolution of Rec 601. It has been superseded by D-MAC.
- **D-MAC** is what we have referred to up to now as Regular MAC because it provides us with regular 700/720 pixel Rec 601 resolution.

- **Wide D-MAC** is the superset of D-MAC with an extended 960 pixel horizontal resolution.
- **HD-MAC** is the higher definition variant of Wide D-MAC which depends on additional digital enhancement signals in order to achieve sharper pictures.

These variations have been designed and developed by European engineers who set their goal on a near perfect picture. There are others, however, who believe that the pursuit of the best can often prove the enemy of the good. These pragmatists recognized that any new TV system which gobbles up two or three equivalent NTSC channels was unlikely to stand on its own feet in the marketplace.

The pragmatists asked the idealists whether the regular D-MAC signal could be squeezed into one terrestrial channel. With great reluctance the idealists were forced to admit that it could be done. If the D-MAC signal was reduced to a half of its normal bandwidth it would still produce a better picture than PAL or NTSC. This new system was first called half MAC or D½ MAC for short. However, with characteristic flair, the marketeers soon dubbed it D2-MAC because that sounded better.

> **D2-MAC** is the bandwidth limited version of D-MAC that has been shoe-horned to fit into one PAL 8 MHz channel. Horizontal definition is thereby reduced from 700 to 450 pixels.

Cheered by this success the pragmatists set another question. There was a good chance that wide-screen TV might herald the next consumer boom in television sales. Could an adequate wide-screen service be provided by squeezing Wide D-MAC into a single terrestrial channel?

Once more the idealists groaned. Yes, it could be done but the pictures would be dreadful. The 450 pixel resolution of D2-MAC would now be spread across the whole of the wide screen, leaving only a meager 320 pixels for the center 4 by 3 picture. The center picture would be no better than NTSC!

In practice, however, the pictures are more than adequate because the MAC coder makes the colors appear "cleaner" than either PAL or NTSC. The accompanying digital CD quality sound is excellent. The two quite separate stereo channels can be used, for example, for simultaneous English and Spanish soundtracks. Best of all the picture is immediately compatible with normal 4 by 3 and wide-screen 16 by 9 screen formats.

Many believe that in our quest for high definition television we have aimed too high too soon. Europe's ten year investment in MAC technology has focused on high end HDTV development and has given little encouragement to the establishment of a simpler wide-screen service. Only the French have dared to risk a major satellite delivered D2-MAC service that offers compatible 16 by 9 and 4 by 3 transmissions. Their wide-screen coverage of major sports events such as the annual Tour de France cycle race and the recent 1994 Winter Olympic Games in Lillehammer is both excellent and free. Perhaps more important, French viewers have demonstrated that they are willing to pay the equivalent of fifty US dollars a month to watch four encrypted wide-screen movie channels, all with CD quality stereo sound.

Has wide-screen D2-MAC come too late?

Like all good thrillers, the story does not end here. There are more important clues to be found.

## Semaphore man

I remember that after the descriptive pages of how to tie knots, which was so easy for my troop leader, my scouting manual contained an illustration of matchstick men with flags. I learned that Semaphore man was always prepared, for he always carried two flags with him with which he could signal his colleague many miles away. My scouting manual must have been a best seller because nearly all modern digital transmission systems have drawn on the semaphore signaling protocol.

Flag semaphore is based on eight possible flag positions which, on a clock face, would correspond to the position of the hour hand at 12:00, 1:30, 3:00, 4:30, etc.

Classical semaphore could easily have been designed with fewer flag positions—such as four (set at 12 o'clock, 3 o'clock, 6 o'clock and 9 o'clock) or with more positions—such as 12 (set at 12, 1, 2, 3 o'clock, etc.).

We would expect that the more rugged four-position semaphore signaling system could be seen more easily than its eight-position brother, but a sequence of two or three flag positions might then be needed to convey just one letter of the alphabet. On the other hand, a 12-position semaphore could convey information 50 percent faster than the eight-position protocol but would need sharper eyesight on our part if recognition errors are to be avoided.

Digital communications systems have been able to improve upon the semaphore flag-waving technique by the introduction of a metronome at both the transmitter and the receiver. In effect Semaphore man no longer has any time to stop and scratch his nose between symbols, for he must produce a valid flag symbol once a second for example, on the dot, without fail.

The introduction of the one second reference tick is important. It means that we no longer need to worry about interpreting those indeterminate symbols which may occur when his arms become tired and the flags then sag a little. We know that on the "tick" the signaled letter will be valid. This eliminates a lot of doubt in our minds and hence makes the communications link much more reliable. This is illustrated in Figure 1-6.

**Figure 1-6**  Precise timekeeping makes the semaphore flags easier to decode.

The metronomic "tick" technique is thus termed *synchronous communication.*

If he is fast, Semaphore man might still have time to scratch his nose between ticks. But we can put even this spare time to good use if we take the clock-face analogy more literally.

So, let us next invest in a stopwatch that has an extra hand that sweeps the whole clock face every second. We can now measure split seconds with some accuracy. Semaphore man no longer needs to wave his arms in all directions. All he now needs to do is to raise a flag at the right time—an eighth or a half a second, say, after the tick. When we check our stopwatch we notice that Semaphore man is an eighth or a half a second late in raising the flag. This deliberate delay on his part can convey exactly the same meaning that we previously understood when he held his flag steady in the 1:30 or the 6 o'clock position.

- Conveying information by raising the flag a little earlier or later than expected is termed *phase shift modulation.*
- The modern day equivalent of telling Semaphore man to raise the flag is called *keying*, from the action of pressing the Morse key.

Communication engineers term this time displacement technique *Phase shift keying* or PSK for short. There is a whole family of such techniques for, just as we had a choice of many semaphore flag protocols (four-position, eight-position etc.), we can design a similar range of PSK protocols.

- If we use only two flag positions on the clock face—at 12 and 6 o'clock—the technique is termed Binary PSK or BPSK for short.
- If we use four flag positions, by including the 3 and 9 o'clock positions, then the technique is termed Quad PSK or QPSK.
- For more complex systems the classical Latin overtones are left behind: 8-ary PSK and 16-ary PSK are exactly what we would expect from an 8 and 16 flag position system.

In practice even more complex coding systems can be used which can combine the concepts of flag timing and flag position.

If we have lots of messages to send we can keep Semaphore man very busy indeed. Sooner or later we may get even busier and may be faced with a backlog of messages. Semaphore man is already working flat out. What do we do?

Should we ask Semaphore man to wave his arms more quickly or should we add more flag positions to the signaling protocol?

There is an alternative.

We could hire one or more extra Semaphore men to work alongside the first. We must take care, however, either to give them enough room or to arrange the synchronization of their arm movements in order to prevent them from inadvertently hitting one another. We shall return to this multi-semaphore man technique later in the book.

Why do communications engineers believe that PSK is so important?

Phase shift modulation is the workhorse of digital communications. It works where everything else has failed. It works well at a 20 dB signal/noise ratio where normal analog techniques are looking groggy at 40 dB signal/noise. It is the technique that provides the digital waiter with an enormous communications advantage over his analog colleague.

You will recall that we could order our cake from the digital waiter without needing to raise our voice above the hubbub in the restaurant. He could hear the "cake-cake-topping" sequence above all but a crash of dinner plates from the kitchen.

To order the same cake from the analog waiter required us both to raise our voice and to wait for a lull in the background conversation. Let us take a more detailed look at the difference in volume between our analog shout and our digital whisper.

**Power to your elbow**

In both the UK and the USA it is easy to be drawn into a conversation about the horsepower of the automobiles we would like to drive one day. We know, from experience, that a 125 brake horsepower compact does not accelerate very fast whereas a turbo-charged 250 horsepower sedan is much more fun to drive.

In Canada and in continental Europe the power of the horse has been replaced by the environmentally cleaner kilowatt (kW) so we should not be surprised to overhear a similar comparison between the 100 kW compact and the 200 kW sports coupé.

But, in both cases, we know that doubling the engine power does not double the top speed of the car. If only it did!

To double the top speed we need to increase the power of the engine much further—500 horsepower or 400 kilowatts under the hood should serve to frighten our passengers quite nicely.

The same rule applies to the output voltage and electric power that we might obtain from a dynamo that powers the lights on a pedal cycle.

As an experiment we can turn the bicycle upside down and use our hand to turn the pedals slowly. If we connect a voltmeter to the terminals of the dynamo we might measure 2 volts. If we use our hand to turn the pedal faster we might see the needle of the voltmeter rise from 2 to 4 volts. The voltage may have doubled but the power, kindly provided by the muscles in our hand and forearm, has risen fourfold. Pedaling faster certainly takes much more effort.

If we had the strength to pedal very much faster we might see the needle move up from 2 to 20 volts, an increase of 10 times. To achieve this increase in voltage the supplied power would need to rise not ten but a hundredfold, a very hard task indeed.

In the earlier section "The Super-Truth in the Snow" we learned that the digital waiter had a 20 dB advantage over the analog waiter (see page 26). Not much really—a ratio of ten times in voltage terms.

But look again: 10 times in voltage terms is a 100 times in terms of power.

- In terms of horsepower, one athlete on a "digital" bicycle can outperform the powerful "analog" turbo sports coupé.
- More important is that, in theory, a 50 kW digital television transmitter can cover the same service area as a conventional 5,000 kW analog TV transmitter.

OK, if this digital transmission technology is such good news why is it not in everyday use?

At last it is.

Ten years ago the success of the United Kingdom closed captioning teletext service stimulated interest in the further development of digital broadcasting. We have seen how the normally hidden teletext area can be expanded in size but how this can normally only occur at the expense of the visible picture.

Are there any other solutions?

One alternative method of sending more information is to try to raise the speed of the teletext data burst from 7 Mb/s to something higher. There is certainly some room for improvement because the basic teletext data waveform was originally designed as a simple "two level" analog system, with flag positions at 12 and 6 o'clock.

The first improvement to be tried was to add in the intermediate 3 and 9 o'clock flag positions so as to convert the data waveform from a two into an analog four level system. Such a technique doubles the data speed from 7 to 14 Mb/s. It was tried and it worked well. Unfortunately, it was too late to apply this upgrade to the established terrestrial teletext service but the idea was immediately included in the design of the new TV standard—MAC TV—which was intended for "direct-to-home" (DTH) satellite TV broadcasting.

As we have seen, the regular MAC TV picture was designed to be sharper than PAL (700 pixels horizontal resolution rather than 420 for PAL and 330 for NTSC) and so would need a wider channel bandwidth for the sharper picture—typically equivalent to two NTSC channels. Why not take advantage of the extra bandwidth needed for the sharper MAC pictures and push up the data rate even further, from 14 to 20 Mb/s. Once again it was tried and it worked well.

The next step was to decide the form of data modulation:

1. This improved prototype four level teletext signaling system could either remain in the analog domain, just like the classical four-flag position Semaphore described earlier or

2. Alternatively the four level signaling system could be

transferred into the "stopwatch" domain by the use of quad phase shift keying (QPSK) in place of the multi-level analog waveform.

There were two schools of thought:

- The "analog" school argued that there was little point in switching over to the more efficient phase shift modulation system (PSK) for the additional teletext service if the main TV picture still needed to be transmitted at high power.
- The "digital" school argued that the more rugged and efficient PSK technique was a step in the right direction for the eventual introduction of digital television.

Ten years later, the analog school appears to have won for there are now four analog MAC TV systems.

B-MAC which uses a data burst at 14 Mb/s
D-MAC which uses a data burst at 20 Mb/s
D2-MAC which halves the data speed to 10 Mb/s
HD-MAC which restores the data speed to 20 Mb/s

Only the baby brother D2-MAC signal with its 10 Mb/s data speed can travel through a European cable TV system without re-engineering or undue difficulty. Its 20 Mb/s bigger brothers need nearly twice as much bandwidth if they are to be distributed through cable TV systems.

The only proposal for a high efficiency PSK MAC TV system, C-MAC, was overturned by analog D-MAC which has, in turn, grown into HD-MAC. But efficient digital transmission did not quite die.

### Digital stereo television sound

The idea of transmitting data at power levels which are a hundredfold below conventional analog techniques was too good to ignore for long. If the phase shift modulation technique could not be championed in the C-MAC format then perhaps it could be retro-fitted to the existing PAL television transmissions in a different way. After all, it seemed possible that a low-power PSK signal might be "hidden" successfully behind a high-power analog signal just as teletext was hidden in the invisible line space above the top of the TV picture.

And so it has proved in practice.

In the UK we have recently witnessed the successful introduction of CD quality digital stereo television sound on all of our four terrestrial TV channels. It has been provided by the additional transmission of a low-power PSK signal which has been neatly hidden away near the edge of the TV channel.

It travels unnoticed down cable TV systems and remains unnoticed by all our TV sets, except for those new TV sets and video recorders that are specially equipped to interpret the signal.

This new signal has been inserted at a level which is set at least 20 dB (100 times) below the power of the TV picture. The major 500 kW TV transmitting stations radiate the digital stereo signal at a power level of no more than 5 kW—which is just the number that our calculations led us to expect.

This field proven ability to hide a low-power digital PSK signal behind the high-power analog transmitters has generated a lot of enthusiasm and excitement, especially in the United States. We know that low-power digital signals can be safely broadcast either at the edge of existing TV channels, as in the UK, or in those spaces between the TV channels that are normally left deliberately empty—

On those spare frequencies that the Americans have delightfully named the *taboo channels*.

Let's recap.

When we first compared digitally compressed television to conventional analog television transmission (see page 78) we discovered that we could use a full-power PAL TV transmitter to support a "Singapore-style" teletext-like data rate of 7 Mb/s.

The subsequent introduction of the "stopwatch" low-power PSK technique allows us to upgrade our initial assessment in two ways:

1. By switching from analog to phase modulation we can drop the power requirement of the transmitter a hundredfold.
2. By increasing the possible number of "flag positions" we can double—or even quadruple—the data rate without the need to increase the channel bandwidth.

These improvements in the means of transmission provide digitally compressed television, or digital television for short, with a number of unique selling points.

Low-power digital technology allows us to set up new transmitters that can broadcast in the previously unused frequencies which have always lain fallow between the existing high-power television channel assignments, on the taboo channels.

The previous dearth of additional terrestrial TV channels, which drove many of us to seek variety from satellite television, is now no more than a bad dream for we can suddenly tune in to the taboo channels and enjoy many more TV programs through our existing rooftop TV antenna.

A digital decoder box between the existing rooftop antenna and the existing television set will provide the extra channels.

We have seen that a 7 Mb/s data stream can provide a single Rec 601 700/720 pixel resolution TV picture. Doubling the data rate to 14 Mb/s would allow us to transmit two such pictures simultaneously through a standard terrestrial TV channel.

Two pictures for the price of one!

But digital television offers something even more valuable for we are not restricted to the choice of just two Rec 601 quality TV pictures.

We are free to "cut" the cake—the 14 Mb/s data stream—in any way we choose.

Digital television offers us **choice.** For example:

- The same 14 Mb/s data stream can provide us with **four** NTSC quality pictures or
- The same data stream can provide us with **five** VHS quality pictures or
- The same data stream can provide us with **eight** pre-recorded pictures from a central video CD jukebox or
- The same data stream can provide us with **ten** or more C-Span pictures from the Congressional committee rooms or
- The same data stream can provide us with **one** high definition quality picture

It is refreshing to know that we are not required to make up our mind as to the "best" choice in this list, just as we are not expected to commit ourselves, irrevocably, to the choice of any one particular daily newspaper.

Digital television technology offers us the unbridled luxury of ducking any decision-making on what we should choose from this list of options, for we shall probably be offered all of them at some time.

Digital television technology will allow the broadcaster to offer a mix of different quality/quantity options. These will occur across different channels and/or at different times of day.

On some occasions many of us may enjoy watching a sporting event such as the lawn tennis tournament from Wimbledon. It would be most welcome if the program could be transmitted in a "high definition" format which makes it that much easier to follow the ball.

Another TV channel might decide, for example, to offer the dedicated band of Congress-watchers the opportunity to eavesdrop simultaneously on all the US committees as well as being able to follow the proceedings from European parliaments.

It need not be expensive. At present the major broadcast networks enjoy permanent "feeds" from the US Congress and the UK Parliament but the signals get no further than the studio mixing desk, for there is rarely any space in the "schedules" to broadcast them any further.

- Digital television places the choice of both quality and quantity firmly in the hands of the TV scheduler on whom the whole success of the channel has always depended.
- Providing that the new technology can support a mix, it would seem more reasonable to assign the decision of quality versus quantity to market forces rather than to the attempted strict enforcement of technical edicts.

In Europe it has been the lack of this marketing flexibility—with the consequent false need to choose between adequate PAL picture quantity or more-than-adequate MAC picture quality, that has made the MAC debate such a stale and acrimonious business. In the final chapter, entitled "Simply Irresistible," we will discuss these programming choices in more detail.

## Full circle

You may just recall that, at the start of this chapter, I set out to buy a new compact disc player. I discovered that the very accurate timing in the replay circuits of the expensive CD player would allow me to hear much finer detail, even to the sound of the musicians shuffling their feet.

In the early days of MAC the development engineers encountered similar timing constraints when trying to add digital sound to the TV signal. The first member of the MAC TV family to reach the commercial marketplace—B-MAC—used a 14 Mb/s data burst to convey both the teletext signals and, more interestingly, to convey a number of channels of compressed digital stereo sound.

The sound compression algorithm, chosen for this purpose, was a differential coding scheme which was licensed from Dolby Labs. Although it is very effective in operation the trained ear can, after a while, just notice a slight impairment in sound quality when compared to the originating 16-bit compact disc recording.

Since then many studies have been conducted on other data compression algorithms to see whether they could approach the sound quality of the 16-bit CD recording more closely. In Europe it was finally agreed that the middle aged members of the MAC TV family—C-MAC and the many variants of D-MAC—would adopt an audio data reduction technique which, in the event, did not rely on differential coding. This technique is called Near Instantaneous Companding Audio Multiplexing—or NICAM for short.

At the heart of NICAM is a smart robot.

A thousand times a second the transmitting robot listens to the music and decides whether it is too loud or too quiet. If the music is too loud the robot turns the volume down to half and if it is too quiet the robot turns the volume up to twice its previous value.

If faced with a sudden crescendo or near silence the robot is allowed to turn the volume up or down as many as four times in quick succession. In this way the sound is "compressed" into a narrower volume range and can thus be transmitted with fewer bits per sample. In other words the "headroom" has been intelligently taken away.

At the receiver the transmitting robot enjoys the services of an assistant whose sole task is to reverse the robot's earlier adjustments of the volume. In this way the original dynamic of the music is restored.

Experimental results have shown that the trained ear can differentiate between:

- a standard 16-bit CD stereo recording which is coded at 1.41 Mb/s (this data rate value was derived on page 67).
- the NICAM stereo signal which can be successfully coded at about half this data rate (728 kb/s) and
- the Dolby differential coding system—termed adaptive delta modulation—which, in the B-MAC system, is usually transmitted at a data rate of about 500 kb/s per stereo pair.

In conducting these experiments it was, of course, necessary to make sure that all the musical test signals were played through near identical high quality loudspeakers, audio amplifiers and digital circuits. In particular it was found that any timing discontinuities in the digital replay circuits caused unwanted "coloration" of the music.

The myth was true—the quality of the digital circuits really did influence the reproduction quality of the music. We are back to the digital bakery and the need for a steady hand in gluing the pieces of cake back together.

The development engineers were able to draw on the results of previous CCIR studies into the effects of transmission impairments to digital telecommunications sound circuits.

> It had been found that, even if the digital samples of the music were set in place to an accuracy of 200 nanoseconds (a fifth of a millionth of a second) half the test listeners would still report coloration.

The CCIR recommended that, in order to avoid coloration—the "thickening" of the sound—the samples should be set in place to an accuracy of better than 50 nanoseconds (a twentieth of a millionth of a second). Even at that point one in twenty of the population might just still notice that the music is not as clear as it might be.

It is perhaps reassuring to learn that we can afford to be a little less rigorous in the actual transmission of the digital signal from studio to home, as what finally counts is not so much the content but the quality of the presentation.

A not unfamiliar theme!

We may finally draw a few more threads together.

- If our ear can notice the minute perturbations in time that are no more than a tiny fraction of a millionth of a second no wonder we need a stable environment in which to fully appreciate the musical arts.
- If we need to synchronize our internal "stopwatch" to such a high precision no wonder it takes us a few seconds to align our thoughts, to establish our concentration as we gently "slip" into the music.
- If we can guarantee that the reproduction quality of the music is as pure as clear spring water, with no coloration or thickening of the notes, then we can see right to the bottom of the

deep pool, right down to where the musicians are shuffling their feet.

If our ear can offer us such transports of delight what might our eye offer? The next chapter, "Perceptions," sets out to shed some light on what our eyes, not our ears, can tell us.

# 2

# Perceptions

## INTRODUCTION

From time to time it does seem that Banks and Television are not quite made for one another.

We have all occasionally enjoyed the "crime-buster" late night TV programs which are introduced by the local police department. The program will often include short clips of prospective bank robbers in action as recorded by the bank's hidden security cameras.

I have often thought that the picture quality is appalling as it does seem nearly impossible for anyone to identify the robber from his televised performance.

The bank's security cameras may be hidden but their TV monitors are not. The TV displays are mounted on sturdy wall brackets in good view of all the customers. These displays are expensive, high quality products which are known to work well.

But there must be a hidden flaw for why else are they now all turned off?

We know that too.

The pictures they once showed of smiling financial advisers and happy, contented customers were also quite appalling.

I remember that the pictures were no more than a pale and flickering imitation of real life which made the real customer in the line, me, turn away from such unpleasant visual intrusions.

- Why were all these television experiments such a turn-off, and will high definition television (HDTV) offer us anything better?
- If we could just start again at the beginning: What would make one TV display both easier on the eye and more compelling to watch than its rival?
- Are there any clues to improving the watchability of the TV display?

Yes, there is hope. It is certainly worth looking over the garden fence at the improvements that have taken place over the last few years in the quality of computer terminal visual displays.

Twenty years ago the quality of computer visual display was no better than today's television sets; after all, they both used the same technology.

There were complaints.

Visual Display Terminal (VDT) operators gradually came to realize that sitting in front of a flickering screen all day long did not seem to do their health a lot of good. Many labor organizations lobbied hard, both for improvements in working conditions and for a better scientific and medical understanding of what made some VDT operators feel unwell.

Studies showed that there was no more than a weak causal relationship between ill health and the workplace use of VDTs. It was generally agreed, however, that it would be helpful if workplace flicker could be reduced or eliminated.

Two main sources of flicker were identified:

- The first source of flicker came, not surprisingly, from the VDT screen itself, and
- The second source of flicker came from the ambient lighting in the room.

I was interested in the medical evidence quoted in UK Parliamentary reports which suggested that, after a little time, our eye and brain can generally "adapt" and nullify the perception of flicker. It was generally agreed, however, that this is a quite needless and non-productive use of brain power which can easily lead to headaches or eyestrain when we are not otherwise feeling at our best.

Using our brain to nullify flicker seems akin to driving a car with the parking brake still applied—it can be done but is a pointless waste of resources.

It is therefore clearly better practice, and much easier on the brain, if background flicker can be reduced. This can quite easily be achieved by the introduction of some flicker-free natural daylight and/or the use of old fashioned flicker-free incandescent lamps in our working environment. This is illustrated in Figure 2-1.

The ubiquitous lines of standard office ceiling strip lights, which use standard fluorescent tubes, are by no means flicker-free. They radiate a very unsteady light, especially towards the end of their working life. Fortunately it is relatively easy to modify these strip lights (by changing the regular coil in the lamp holder for a "high frequency" unit) in order to reduce substantially the amount of flicker that they generate.

But the prime culprit of flicker is the display screen itself. Many studies therefore recommended that the display should use a high refresh rate (the picture renewal rate) to appear "flicker-free."

But how high is high?

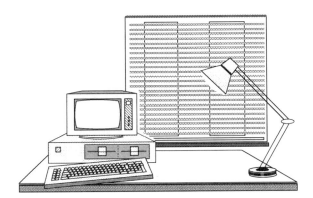

**Figure 2-1**  The level of background illumination should be similar to the brightness of the display screen.

In 1987 the British Standards Institute (BSI) embarked on a set of standards for VDT ergonomics, otherwise called VDT human factors. These were intended to be harmonized with other, similar, standards that were being developed both in the European Community (EC) and elsewhere in the world.

I was particularly intrigued to find out how the BSI would define that most elusive of qualities "flicker-free." I hoped that the definition might possibly just shed some light on why the television set in banks and shopping malls had been such a turn-off. Might this definition perhaps un-earth some firm scientific criteria for what constitutes a good computer display? Could the same criteria be then applied to the selection of a television display which could be used to good effect in retail outlets? Could we even make the display so good that people would no longer turn away but would actually like to watch the screen?

It is then only a small step to apply the same criteria to our television sets at home. We could then argue that if a visual display terminal is such a joy to work with during the working day, surely we deserve nothing less than the same high standard from our television set at home.

Might this quest for the "good" visual display terminal also prove the best route to that most elusive of holy grails? The definition of a satisfactory, domestic quality, high definition TV standard.

The resulting definitions might be quite a surprise. This indirect approach to defining high definition television might even allow us to sidestep today's biggest gamble—the choice of a suitable line standard for high definition television.

As we know only too well, the current American and European HDTV standards seem stunningly naive. Both camps have suggested that:

> Depending on where we live, take the current number of TV
> lines being offered, 525 lines in NTSC or 625 lines in PAL,
> and double it to 1,050 or 1,250 lines. That is HDTV.

The only obvious merit of this American and European approach is that it places a rather expensive each-way bet on whether the Japanese, in their own 1,125 line HDTV system, originally pitched the number of lines either a shade too high or too low.

At first sight the independent observer—an endangered species at the best of times—must therefore assume that, compared to the Japanese 1,125 line system:

- The Americans must believe that the loss of a few lines will make no difference to picture quality, and
- The Europeans must believe that a few more lines will make all the difference to the picture.

Surely we can do better than that.

## FLICKER

It is said that an elephant is notoriously difficult to define but we will recognize one when we see it.

Flicker is much the same. Pages of diligent definition and discussion can easily count for nothing for it is only our eye and brain that can really and finally tell us whether flicker is occurring "before our very eyes."

Despite this elephantine reservation, I shall try, in the course of the next few pages, to shed some light on what makes pictures appear to flicker or, perhaps better still, what makes them appear steady. I shall also develop the idea that the middle ground, finely pitched on that narrow perceptual borderline between "flickery" pictures and "steady" pictures, is an area which is ripe for development.

When the BSI VDT ergonomics team started their work in 1987 they had no choice but to bite the bullet and try to gain a scientific understanding of what made pictures flicker.

It was not easy.

They knew that parts of the VDT standard would, inevitably, be technically demanding so they formed an expert committee— the BSI PSM/39 Applied Ergonomics Technical Committee—to sift through and analyze the available and often conflicting evidence.

I had expressed an interest in their deliberations and so they kindly provided me with copies of both the draft standards and of some of the background material on which they had based their findings. Their first step was to define some parameters:

## Subjectivity

I found their definition of "subjectivity" to be quite tantalizing. Their definition lifts the veil only the merest fraction on what interactions might be taking place between one's eye and one's brain.

### Definition

In practice the detection and recognition of visual images is not solely dependent on the quality of the image on the display. It is also influenced by the degree to which the user understands the communication as a whole and can anticipate or interpret the image.

## Viewing distance

For domestic viewing it has become customary to define the viewing distance from a television picture as placing the viewer at so many "units of picture height" away from the screen. In contrast, the definition of the VDT viewing distance includes a dependence on how the information is being shown on the screen.

### Definition

The distance or range of distances between the screen and the operator's eye for which the images on the screen meet the criteria for character size, resolution, jitter and flicker.

## Luminance balance

Even the brightness of the surroundings is given much more importance for the VDT operator than for the TV viewer at home. The amount of background, or ambient, light in the room helps define the luminance balance.

## Definition

Large differences in the luminance of sequentially viewed task areas, between the screen and an adjoining document for example, can cause a momentary loss of contrast sensitivity and should be avoided.

Sequential luminance differences should not exceed a ratio of 10 to 1 though these should be preferably restricted to a 3 to 1 range. Furthermore, the static background should not differ by more than 20 to 1 from the brightness of the screen.

The minimum light output of the VDT should exceed 35 candelas per square meter ($cdm^{-2}$). It is perhaps worth commenting that this is not very bright by today's standards for, by comparison, conventional domestic TV screens produce illuminances of about 200–300 $cdm^{-2}$.

On the other hand, at the present state of development, the light output of HDTV screens is currently so low that we are usually invited to watch an HDTV demonstration in a darkened room.

The implications of much higher light outputs from display devices are discussed later.

### Jitter

The VDT standard is quite emphatic that the displayed image must be free of jitter. In order to avoid subjective discomfort the anti-jitter requirement is very strict. In numeric terms the BSI recommend that:

## Recommendation

The variations in the geometric location of an individual picture element shall not exceed 0.2 mm per meter of viewing distance when measured over a period of 1 second or less.

In everyday terms this means that, on each refresh of the screen, the raster lines must stick to the same path on the display screen very closely indeed. This is illustrated in Figure 2-2. Unlike

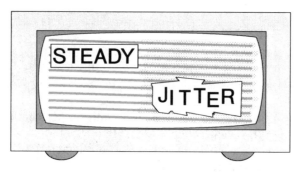

**Figure 2-2**  Jitter makes the displayed image more tiring to view.

watching TV at home, the VDT operator needs to sit quite close to the VDT—no more than an arm's length from the screen. At the usual working distance of 24 inches (60 cm) between the operator and his display terminal the raster lines are not allowed to wander more than a tenth of a raster line from their usual path across and down the screen.

This is a tough requirement but serves to explain why all modern computer displays have adopted sequential (progressive) scanning in preference to an interlaced scan as the most practical means of meeting the jitter requirements.

### Progressive versus interlaced scanning

The traveling spot of light on the face of the normal TV screen—the Cathode Ray Tube—starts its journey at the top of the picture and gradually works its way down to the bottom. Once it reaches the bottom it must rush or "fly back" to the top of the picture to start its journey all over again.

As the spot of light moves relatively slowly down the picture it must also move quickly from side to side—its horizontal motion—in order to fully paint the screen.

We can perform exactly the same maneuver, albeit very slowly, when watering the garden with a garden hose that is fitted with a spray head. We can wet quite a large area of the garden by moving the hand-held spray head back and forth in one direction as we walk in a different direction across the garden.

We can adopt one of two principal gardening strategies:

1. We can walk slowly and give the garden a good soaking. This is similar to progressive scanning.
2. We can walk faster, giving the garden an initial soaking, then return to do the bits that got missed out the first time around. This is similar to interlaced scanning.

It is quite easy to see that if we swing our "watering" arm a hundred times in our short promenade across the garden we will succeed in drawing a hundred wet lines on the ground. Interestingly, this is true of both progressive and interlaced watering techniques. Both methods take the same time to soak the ground, whether done once slowly or twice quickly.

Television sets produce the raster lines in exactly the same way as we can draw wet lines on the ground. To produce "progressive" 525 lines on the screen the back and forth (horizontal) spot driver circuit must operate 525 times faster than the top-to-bottom (vertical) spot driver circuit.

It is as easy as that.

Which scanning method is better, progressive or interlaced?

The answer depends on resources or, if you prefer, on how much time you can afford to spend watering the garden.

Imagine that we have a small garden and that we can afford to set aside one hour a week for doing the watering. We can work progressively around the garden and give all the plants a thoroughly good soaking.

Next imagine that we have suddenly acquired a bigger garden which will now need more watering time than we can afford to set aside for the task. We could properly water one half of the garden one week and complete the other half the following week. There is a high risk, however, that some of the plants will die before we ever reach them. Alternatively we could rush around the whole garden every week, giving the plants barely enough water to get by. Thus, providing we have the (technical) resources, progressive watering is a better technique. When resources are scarce, interlaced watering is much better than nothing at all.

This comparison with watering the garden has thrown some light on the mechanics of progressive and interlaced scanning methods. It has not touched on another important area—what we are able or not able to "see" or perceive in the two different types of picture presentation. For an insight into this subjectivity we can usefully turn to a new analogy—that of reading a book.

### Reading a book

The difference between interlaced and progressive scanning is perhaps not unlike the different ways we might read a book.

- The progressive reader starts his book on page one and continues steadfastly right through to the end.
- The interlaced reader skips pages—he works his way quickly through the book so that he can find out what happens at the end. He then starts again near the beginning and is generally successful in filling in the missing bits of the plot, provided that he uses his brain.
- The interlaced reader gets through books faster, but would probably admit that he gets less from them than the progressive reader who "settles down" and enjoys the plot as it unfolds before him.
- The interlaced reader's brain has to work harder, though this can be fun. Before he can understand the story in full he must first sort the new "even-page" information into its proper place in the skeleton plot that has already been provided by the "odd-pages" which he read earlier.
- The progressive reader is able to avoid these mental gymnastics: the continuity of the plot is assured and all he need do is to sit back and enjoy it.

We know that both the progressive and the interlaced readers readily agree that the progressive reader seems to get "more" from a book than the interlaced reader. But how much more? 10 percent, 50 percent, 100 percent more?

This is clearly a tricky question to answer so let us switch back for a moment to some television scanning experiments where scientific measurements can be made more easily.

Television studies show that, for the same number of scanning lines, progressive scanning offers about 40 to 60 percent

more "perceived definition" than the more conventional interlaced scanning.

We can put this another way by turning the argument on its
head. For the same perceived definition, we need about a third
less TV lines if we choose progressive scanning in favor of interlaced scanning.

In terms of the book analogy: a 100-page book might give as
much pleasure to the progressive reader as an equivalent 150-
page book will give to the page-skipper. Yes, this does sound about
right.

The same arguments can be applied to bigger books.

A 700-page book might give the same pleasure to a progressive reader as a 1,000-page book will provide to the interlaced
reader. We shall return to this 700–1,000 page comparison a little
later on.

Can we draw any other conclusions from our two types of book
reader?

In real life we often alternate between the two styles of
reading. Sometimes we enjoy "progressive" reading, but at other
times we enjoy "interlaced" reading. The style that we choose
usually depends on how we feel at the time.

If we want to delve deeply into a complex subject we appreciate it when we are led gently and progressively along a new
line of thought.

"Full marks for presentation," we might exclaim, "I now
understand it."

The book might be quite short but might uncover a few
"gems" that would otherwise be missed by the page-skipping
interlaced reader. On the other hand, the plot might be so "slow
moving" that the progressive reader might, later on, secretly wish
that he had cheated a little and skipped a few pages.

The television scanning studies bear this out (see bibliography). Fast moving scenes, which are a test of the dynamic
resolution, look better when seen on a progressively scanned 700
line screen than when viewed on a 1,000 line interlaced display.

In slow moving scenes, which are a test of the static resolution, the 700 line progressive scan screen loses some of its advantage to the interlaced 1,000 line pictures.

In order to demonstrate these findings, imagine that our task is to turn the page-skipping interlaced reader into a steady progressive one. We might ask the page skipper to read the book and then tell us the story in his own words. He would read the book in his own fashion, first reading the odd pages right through which would then be followed by the even pages.

If we expect him to retell the story he would need to remember the whole book before he could tell the tale in the correct order.

We can perform the same trick with the interlaced television picture. We can first take all the odd TV lines (the odd field) in the TV picture which are then followed a split second later by the even TV lines (the even field) and store them in an electronic memory. We now have the whole TV picture under our thumb and can read it out in any way and at any speed we wish.

But this is not quite the good news we hoped it would be.

For "still" TV pictures this storage technique works very well but for moving scenes this technique produces TV pictures of the most appalling quality. At first sight it appears that our analogy of the page-skipping interlaced reader has let us down completely.

The reason for the breakdown is plain.

A book is a record of a set of immutable events which have been frozen in time whereas the television picture is a view of life as it is happening. It is as though the book was being rewritten at the same time that the reader was working through the pages.

This rewriting does not cause distress to the progressive reader; a revision of Chapter One will cause him no concern if he has already reached Chapter Three. For the interlaced reader, however, whom we have asked to retell the story, these revisions can cause utter havoc. When he returns to read the even pages of Chapter One he may discover that the plot has moved from England to America and that the secret agent, who appeared unannounced on page 3, has now been killed off on page 2.

The interlaced reader does his best to take these changes in his stride for, by now, he has become quite philosophical. He has learned that we cannot understand everything, life has moved on and we must do our best to catch up as best we can. He would suggest that our best course of action is to ignore the out-of-date happenings on the odd pages and instead concentrate our attention on the more up-to-date even pages.

He could well be right, to ask him to retell the story of the whole book would make less sense than agreeing to discard the odd pages and just tell us what he is able to about the more recent even ones.

The rapidly moving interlaced television picture presents us with the same problem. Storing the whole interlaced picture in an electronic memory leads to a total mix-up of exactly what happened when. There is thus no easy way to switch from interlaced to progressive scan—though the reverse transformation from progressive to interlaced scanning is trivially easy.

We should be grateful that our eye-brain allows us to deal with rapidly moving events more intelligently than electronic circuits. We are able to discriminate between newer and older information that is separated by no more than a split second on the screen. In effect our brain is trading off picture definition for a partial understanding of what is happening. It is therefore no surprise to learn that, in subjective tests of fast moving scenes, a 1,000 line interlaced picture appears to have no more than 500 lines of resolution. Our brain has taken the only sensible course of action available and has thrown half the pages away!

### The pupil of the eye

Everyday experience has taught us that the pupil of the eye varies in size with the brightness of the scene. In sunlight the pupil diameter is quite small and in a darkened room, such as we experience inside a cinema, it gets much larger.

Studies have shown that we may expect a quite wide variation in the size of different people's pupils when the measurements are made in dark rooms but there is comparatively little variation between different people's pupils when our eyes are measured in sunlight.

The eye seems to work best over the middle range of illumination and pupil size. Objects normally appear to be sharper the better they are illuminated. Hence the welcome improvement offered by a strong reading light when our eyesight is getting past its best.

As the light gets stronger the pupil gets smaller and the focus gets better, just like taking photographs through the lens in a camera. But there is an upper limit to the increase that can be gained in performance by "stopping down" the lens. Above a certain brightness, and hence below a certain minimum size of pupil, there is no further improvement in sharpness. In its place there is another less pleasant effect, which has often been reported by late night revelers—their eyes just hurt from the unexpected strength of the light of day.

In bright sunshine sunglasses can provide a welcome "overdrive" option for our eyes, relieving and easing the workload on the muscles that regulate the size of the pupil. If the eye muscles are likened to the engine of a car we have allowed them, through the use of sunglasses, to operate at a less demanding "rpm."

When we watch television, the size of the pupil of our eye varies according to both the illumination of the viewed screen and the background illumination of the room. Dimly lit scenes will elicit large pupil sizes and bright screens will cause our pupils to contract.

In 1972 Davson reported on a series of earlier experiments in which measurements of pupil diameter were taken from a number of people with normal eyesight (see bibliography). It is interesting to note that the results indicate that, once the illuminance of the screen rises much above 100 cdm$^{-2}$, there is no longer a wide variation in pupil size between different people.

As we can see from Figure 2-3 the size of our pupil will contract to a diameter of about 2 mm at very high (bright sunshine) levels of illuminance. It seems unfortunate that diffraction effects within the eye counteract any further increase in visual acuity that an even smaller pupil diameter might have offered at these high levels of retinal illuminance.

Perhaps "evolution" knew that sunglasses were not far away!

In comparison to Nature's enormous night-day change in light levels, our domestic television sets cover only a narrow range

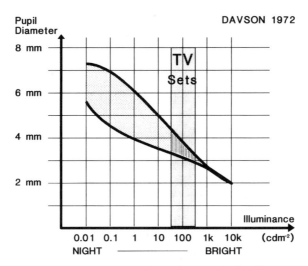

**Figure 2–3** Experimental results—changes in the diameter of the pupil of the eye with ambient illuminance.

of illuminance. We can conveniently subdivide this range into two adjoining, or contiguous, areas.

| | |
|---|---|
| The lower TV illuminance range of | $30\text{–}100 \text{ cdm}^{-2}$ |
| The higher TV illuminance range of | $100\text{–}300 \text{ cdm}^{-2}$ |

Early color television sets and today's High Definition TV displays fall into the lower illuminance range. Modern conventional TV sets can achieve screen illuminances of $300\text{–}400 \text{ cdm}^{-2}$ with the prospect of even higher brightness yet to come. But, for some manufacturers, it would appear that today's screens are already too bright. Is it the imminent onset of flicker or is it purely the vagaries of modern television set styling that now leads many set makers to hide the television display tube behind tinted glass?

We may expect a wide statistical variation in pupil diameter under conditions of low TV set screen illuminance as experimental results indicate a 25 percent inter-individual variation at an illumination of $30 \text{ cdm}^{-2}$.

These results also suggest that we should not be surprised by widely differing reports of people's perception of today's relatively low brightness High Definition TV displays. On the other hand, as the technology of High Definition displays improves and the screens become much brighter we should expect considerably less variation in perception.

### Critical Flicker Fusion

Our own experience seems to suggest that TV displays which are very bright and with a low screen refresh frequency are very likely to exhibit flicker whereas displays with low brightness and high refresh frequency are likely to appear steady.

Somewhere between these two extremes we hope and expect that the flickery image will somehow "fuse" into a steady one. We are not quite sure exactly where this will occur as there does not seem to be any obvious change-over point.

Experimental physiologists have kindly called this the point of "Critical Flicker Fusion." They have discovered that this is not just one point but is a whole series of points—in fact a whole series of graphs.

This, in itself, is enough to strike fear into most hearts so let us press on slowly.

Let us first recap on what we know:

1. A dim TV display usually looks "steady," whereas a very bright TV display can appear "flickery" on some scenes.
2. An old-fashioned home movie cine projector, running at 16 or 24 frames per second, produces "flickery" pictures whereas the domestic TV set at home seems to produce much steadier pictures.

We can therefore say, with some confidence, that as technical improvements continue to make the television display even brighter, at some point we will consequently need to increase the refresh frequency in order to avoid falling into the flicker trap.

There is a more or less direct relationship between the screen brightness and the lowest refresh frequency needed to avoid flicker.

For reasons that are long since lost in the annals of time, North America and Europe adopted differing frequencies for the supply of alternating current (AC) electricity for our homes and offices. The Europeans chose a supply frequency of 50 cycles per second—now called Hertz or Hz for short. The Americans chose the slightly higher electricity supply frequency of 60 Hz.

When electronic television was first developed in the 1930s the early pioneers had little choice but to adopt these frequencies—50 Hz in Europe and 60 Hz in North America—for refreshing the display. The same refresh frequency, whether 50 or 60 Hz, was used throughout the whole television production chain, right from the studio camera through to the domestic television set.

Bar a small (0.01 percent) change in the American TV specification when color television was first introduced in the early 1950s these two separate screen refresh frequencies remain in use to this day—50 Hz in Europe and 59.94 Hz in the USA.

> These two separate production and distribution standards have been, and remain, a major millstone around the neck of every television company which wishes to export its product throughout the world.

One extremely small piece of good news, which we can set against all this gloom, is that it is relatively easy for us to confirm the validity of the graph shown in Figure 2-4.

The graph is a plot of TV screen brightness against TV screen refresh frequency. The plot shows a diagonal fence, marked "Flicker Threshold," which separates the flicker and non-flicker areas.

Three values of TV screen brightness are shown—"Low, Medium and High" though their exact values have, for the moment, been left deliberately vague. Along the bottom axis is shown a European 50 Hz television, an American 60 Hz television and a Visual Display Terminal which operates at the recently agreed 72 Hz refresh standard.

The graph tells us what happens when we experiment with three equally bright Cathode Ray Tube Displays—a European TV set, an American TV set and a Visual Display Terminal—which are set up alongside one another.

*Step One.* We set the brightness control of all three displays to "Low." The score of three ticks on the graph shows that all three displays are perceived to be free of flicker.

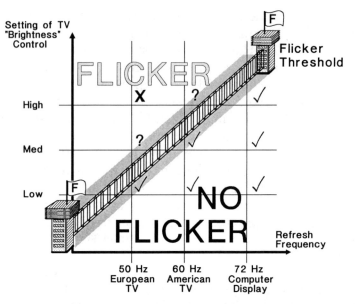

**Figure 2-4**  Flicker threshold—the diagonal frontier

*Step Two.* We turn up the brightness control of all three displays to "Medium." There are now only two ticks in the score. The European TV set is on the borderline of exhibiting flicker, hence the "?"

If we were now to either turn down the brightness or turn up the refresh frequency this will place the European TV the other side of the diagonal Flicker Threshold line. Either change will eliminate the flicker.

*Step Three.* Finally we turn up the brightness to "High." The European TV set scores an "X" as its flicker is now quite awful. The American TV set scores a "?" as it is now on the flicker borderline. Only the Visual Display Terminal is still flicker free.

This is only a simple experiment and has a limitation. It only measures our short-term judgment of flicker which, as we shall discover in a moment, is more exacting than our long-term judgment.

Americans who visit Europe often comment on how unbelievably "flickery" they find the European television sets. How can the Europeans put up with such flickery pictures? Yet, after a few weeks of acclimatization, they no longer appear to perceive the European television flicker. Why or how does this short-term flicker effect wear off?

You will recall the medical evidence that was quoted in the UK Parliament's report on "Visual Display Units" (see bibliography). It suggested that our brain will, after a while, adapt itself to flicker but that, in doing so, this process needlessly uses up a lot of our "brain power."

A good analogy might be that of trying to set up a personal computer (PC) to operate with a number of "background" programs which, we have been assured, will occasionally come in useful. When we have installed all these programs the computer still works, but is now rather slow and may sometimes "crash" unexpectedly.

> Might we therefore tentatively ask whether the running of a cerebral "background" 50 Hz flicker elimination program when we visit Europe places us at an intellectual and competitive disadvantage to our 60 Hz cousins who stayed at home?

The Flicker/No Flicker graph which we discussed overleaf was quite specific about the values of the refresh frequencies of the television sets but remained rather vague about the illuminance of the screens. The "Low," "Medium" and "High" settings of a TV set brightness control offer a qualitative feel to what is happening but the serious experimenter would appreciate something more specific.

We are only too aware that flicker is an entirely subjective phenomenon. We therefore need to aim at a measure of our subjective "perceived illuminance" of the display screen, rather than just a straightforward (objective) measurement of screen illuminance, which we could easily obtain from a photographic light exposure meter.

### Nighttime trolls and Trolands

The next important step in determining whether or not we see flicker is to check the size of the pupil of the eye as this regulates the amount of light falling on the retina. The way in which our pupils contract in bright light is well documented so we can now examine how this affects our perception of flicker.

When we drive through a residential area in the late evening, we cannot fail to notice that our neighbors seem to have bought television sets that flicker much more than our own TV set at home. What has happened, of course, is that our eyes have adjusted to the darkness of the night and the size of our pupil is now bigger than when indoors. The result is that more light than usual enters the eye from each neighbor's TV set than from our own TV set at home.

If we refer back a few pages and consult the graph of pupil size measured against light levels, Figure 2-3, it seems probable that, in going out in the dark, the pupil diameter has widened from about 3 mm to 6 mm. In more familiar terms the pupil has widened from an eighth to a quarter of an inch in diameter.

Thus a simple doubling in the size of the pupil of our eye is more than sufficient to breach the flicker threshold for a bright domestic television set. This is especially noticeable in Europe where the TV refresh frequency is lower.

Let us recap:

We know that the size of the pupil gets smaller as the surroundings get brighter.

This is a helpful in two ways:

- First, our visual acuity—our eyesight—improves as the TV display gets brighter.
- Second, the smaller pupil size helps reduce our susceptibility to flicker—just the opposite effect of driving at night which was described a moment ago.

So, we can say that:

The larger the diameter of the pupil, the more energy will fall onto the retina at the back of the eye and the more likely are we to perceive flicker.

Can we put this statement on a scientific footing?

The light energy that makes its way through the pupil to reach the retina, aptly termed the Retinal Illumination Energy, is measured by a delightful, almost fairy-tale unit of measurement called a Troland.

Every day our retinas enjoy Trolands-full of light energy—provided of course that the physicists and physiologists take care to measure the area of the pupil in square mm (mm²) and measure the screen luminance in candelas/square meter—cdm⁻². For the mathematician we can say:

> the steady state or DC component $E_{DC}$ of the retinal illuminance is the net screen luminance ($L_{SCREEN}$) times the area of the pupil. It is customary to express this in terms of pupil diameter $d$ so that:
>
> $$E_{DC} = L_{SCREEN} \times \pi(d/2)^2 \text{ Trolands}$$

In more everyday language the light energy reaching the retina is very nearly proportional to the illuminance ($L_{SCREEN}$) of the TV screen. For the purist, however, there are two factors that make this relationship less than perfect.

1. When light energy is low, at night for example, our pupils get bigger in order to collect as much of the scarce light energy as possible.
2. When light energy is low there is also a significantly wider spread in the size of pupil between different people.

Trolands are clearly quite delightful but we still need some more clues in our search for an understanding of flicker.


## THE HAMMER BLOW

Some years ago Do-It-Yourself (DIY) home improvements suddenly became much easier. Virtually overnight the hammer drill became widely available in DIY stores and the budding home improver was able to abandon any lingering fear of drilling large, deep and, hopefully, accurately placed holes in his walls.

> Captain Kirk would be delighted: the high speed percussive effect of the DIY hammer drill has made holes appear in places where no holes have appeared before.

Just like our unsuspecting walls at home, our biological systems can also fall easy prey to a wide variety of percussive effects. During hospital operations, surgeons routinely use tightly focused radio energy in order to stop bleeding and to cut tissue. These techniques, variously called electro-surgery or diathermy, can use either continuous or pulsed radio power. Switching across from continuous to pulsed radio power offers the surgeon the same increase in performance as we enjoy when selecting the hammer action on the DIY electric drill at home.

The surgeon, just like the DIY enthusiast, is delighted with the hammer blow action of pulsed radio power. The hospital technician, however, would rather have the surgeon switch back to using continuous power—as the pulses of radio power have an uncanny ability to interfere with much of the patient monitoring equipment (see bibliography).

Our biological affinity to hammer blows can be a mixed blessing.

Research has yet to determine whether or not our bodies will be unduly susceptible to the pulsed output power of the new generation of digital cordless telephones.

However, industrial processes which generate continual vibration or percussive effects on one's hands or limbs are known to lead to long-term injuries such as "white finger" or can damage the tendons in one's wrist or forearm.

Such hammer blow or repetitive strain injuries (RSI) are not uncommon among typists or operators of the keyboard of visual display terminals. My neighbor, who is a local physician, reports a number of cases of repetitive strain injury among dedicated music students.

I am told that pianoforte students are most at risk. Half the students at one London music school were reported to be unable to complete their course of study because they had fallen prey to repetitive strain wrist injury caused by too many long hours of dedicated piano practice.

If one's hands, wrists and internal organs can be so susceptible to mechanical and radio frequency hammer blows, it can

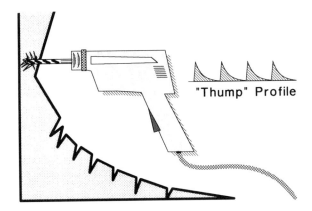

**Figure 2-5**  The percussive "thump" effect of the hammer drill.

come as no surprise that one's eye also notices percussive effects.

The magnitude of the optical percussive effect, which we usually perceive as flicker, is dependent on both the amount of light energy reaching the retina and the profile of the hammer blow itself. We are already familiar with the first of these two parameters—the steady state retinal light energy $E_{DC}$ as it is measured by that fairy-tale unit, the Troland.

We now need to define a new "thump" parameter which can give us a measure of the effect of the hammer blow upon the retina.

### Thump, thump, thump

During the passage of the 1990 UK Broadcasting legislation through the upper parliamentary chamber, the House of Lords, the government introduced an amendment which sought to define "pop" music. The Home Office Minister, Earl Ferrers, read the new, long-winded definition of pop music to the House and then apologized for its length as follows:

The somewhat intricate style of the amendment was thought appropriate because the parliamentary draftsman thought that my definition of "pop," which was thump, thump, thump, would not be parliamentary or statutorily adequate.

*Column 752 Hansard (Lords) 16 Oct. 1990.*

Earl Ferrers' "thump, thump, thump" has two easily understood characteristics.

The first is the rhythmic, repetitious element which seems to appeal to all music lovers, whether they are devotees of classical, jazz or pop.

The second characteristic is more subtle: it is the actual make-up of each individual thump. Audio scientists have analyzed such sounds. They talk of the "attack time" and "decay time" of sounds and can thus explain why an orchestra can sound very different when it is moved to a new concert hall.

We can define optical thump in the same terms that we apply to aural thump.

The profile of the light which is emitted from a TV cathode ray tube, a CRT, is merely a much faster version of a familiar event—the light profile of a firework at an evening's fireworks display. At such a display a firework will suddenly burst into brilliant life above our heads and then gradually fade away. We would describe the "attack time" of the light output of the firework as very fast and its "decay time" as very slow.

In a cathode ray tube the attack time of the light output is fixed and it is always virtually instantaneous. However, we are able to choose a range of different decay times by choosing a range of different light emitters—different phosphors—which are then thinly deposited inside the transparent front face of the cathode ray tube. From an infinitely wide range of possible decay times, these days we only need specify one of three types—long, medium and short persistence phosphors.

- Long persistence phosphors are used in the classical rotating radar display in an effort to keep the picture visible on the screen for as long as possible.
- Medium persistence phosphors are used in our domestic TV sets, and

- Short persistence phosphor tubes are only used for special applications such as in television telecine machines.

Physiologists have researched the optical thump effect quite thoroughly. In 1983 Oppenheim & Willsky (see bibliography) derived a thump formula tailored specifically for cathode ray tubes. The formula looks complicated but the results are straightforward.

$$\text{Amp}(f) = 2/[1 + (\alpha f)^2]^{1/2}$$

Where Amp($f$) is the thump factor, $f$ is the refresh frequency of the CRT display and is the phosphor decay constant.

In practice the expression $(\alpha f)^2$ is always very much smaller than unity so the value of the thump factor, Amp($f$), is always very close to its limiting value of 2. [Even at the onset of picture smudging, which we can take as occurring at a phosphor decay constant of $= 1/5f$ for a 1 percent smudge, the value of Amp($f$) has only fallen 2 percent to 1.96 from its limiting value of 2.]

Thus, as promised, the result is straightforward:

Cathode ray tube displays have a thump factor of 2.

This result is, however, only true for CRTs. Other forms of display, such as Liquid Crystal Display panels will usually exhibit a smaller thump factor.

Let us return to the hammer drill analogy for a moment. The performance of the hammer drill is clearly a product of the continuous, steady power of its electric motor and the thump factor of its hammer mechanism.

It is exactly the same for the eye.

We have already worked out the steady state Retinal Illumination Energy $E_{DC}$ (in Trolands) and we now know the thump factor for the TV CRT display: it always has a value of 2.

Put another way, the effect of the steady state Retinal Illumination Energy on our perception is always multiplied by a factor of 2 when we watch today's TV.

In mathematical terms we can say that:

$$E_{AC} = 2 \times E_{DC} \text{ Trolands}$$

## REFRESH FREQUENCY

We are nearly home and dry but there is still one major element missing from our calculations—the refresh frequency.

So far we have measured the illuminance of the TV screen and then measured the pupil diameter of the eye. We can then indulge in a little arithmetic and obtain a value for $E_{AC}$.

A moment ago we very nearly succeeded in trapping the refresh frequency parameter in our definition of the thump parameter. Unfortunately, despite our best efforts, the value of the CRT thump stuck rigidly to 2 whatever the refresh frequency of the display.

We need to try again.

We clearly need some more experimental data which we might then fit into a form similar to the earlier Flicker/No Flicker graph of Figure 2-4. You may recall that this graph had the vertical axis marked "TV set brightness control" which, as we agreed at the time, seemed rather vague. We now have a chance to substitute values of Perceptive Trolands—$E_{AC}$—in its place. We can, in fact, go ahead and do this provided we also take care with the other axis, the Display Refresh Frequency axis.

We should avoid using domestic TV sets or Visual Display Units for the experiments, instead we should use a simple light projector and plain white screens which do not use raster lines.

We should not use pulsatile light or otherwise we will need to calculate and apply an optical thump correction factor. Instead we should use sinusoidal light, ordinary AC electric light, which has an intrinsic thump factor of unity, a value of 1.

The experiment must be kept basic and simple, so that there are as few things to go wrong as possible. We do, however, still need to make one further decision.

How big should we make the viewing screen?

Here we hit another snag. Our perception of the size of the viewing screen is entirely dependent on how far away we sit from the screen. It would be better if we could choose a parameter that combines both the size of the screen and the viewing distance.

The idea of a "Viewing Angle" is a perfect combination of

screen size and viewing distance. The viewing angle is usually expressed in one of two ways.

Picture height    The viewer is seated so many units of "picture height" away from the screen.

Subtended angle   More scientifically we can fit a student's geometry protractor to Figure 2-6 and directly read off the viewing angle.

The television industry quote viewing distance in units of picture height whereas scientists prefer to quote subtended viewing angles as they assume the experimental screen is not rectangular but *circular*.

We now have enough information to conduct our critical flicker fusion experiment which is illustrated in Figure 2-7.

We can sit our volunteers in front of the circular screen and ask them to tell us exactly when they both do and do not see flicker. We can change the illuminance—the unit brightness—of the screen and we can also change the frequency of the electric power supplied to the projector lamp. We must also make sure that we keep a check on the size of the volunteers' pupil.

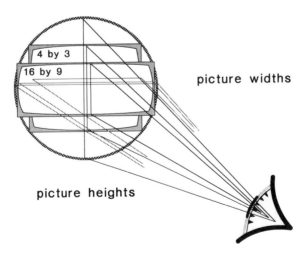

**Figure 2-6**   The relation between picture width and picture height for regular and wide aspect ratio display screens.

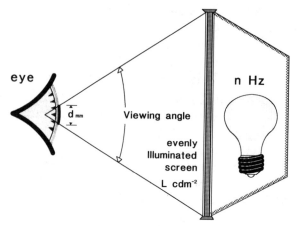

**Figure 2-7** Experiment to determine the critical flicker fusion
frequency.

We can then change the viewing angle, either by positioning
the volunteer closer to or further away from the screen or by
occluding part of the screen so as to make it appear smaller.

We can then rerun the experiment all over again.

We must conduct these experiments hundreds of times so that
we can establish the average—the statistical mean—and the
differences—the statistical variation about the mean—that
exist in the population.

You will be delighted to learn that, as in the best traditions
of television cooking programs, all this hard work has been done
before and the results are awaiting our pleasure below.

The most distinctive feature of the experimental results is
that it shows that our perception of flicker changes markedly
with viewing angle.

The graph in Figure 2-8 thus shows four Critical flicker
fusion (CFF) Threshold curves which were measured at four
viewing angles of 10, 30, 50 and 70 degrees. This is not entirely
unexpected: we have always been aware that our side vision was
more sensitive to flicker than our central vision. What is of
interest is how this differential sensitivity changes with increas-

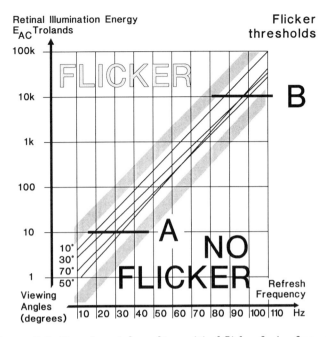

**Figure 2-8**  Experimental results—critical flicker fusion frequency
plotted against retinal illumination energy.

ing levels of Retinal Illumination Energy—$E_{AC}$. At low values of
$E_{AC}$, at 10 Trolands for example, it looks as though our sensitivity
to flicker increases fairly smoothly with widening viewing angle.

But look again. What is shown in line A on the graph in
Figure 2-8 is not what we expect.

Our sensitivity to off-axis flicker reaches a maximum at ±25
degrees off-axis (a total of 50 degrees of view) and then falls
back a little at ±35 degrees off-axis.

At high values of $E_{AC}$ the situation is quite different.

At 10,000 (10 k) Trolands—line B on the graph—the flicker
threshold lines are no longer evenly spaced but are bunched.
The 10 degree viewing angle CFF threshold line, repre-
senting central vision, is now set some way away from the
other three threshold lines.

The results show that, at medium and high levels of illuminance, a **distinct change** takes place in our perception of flicker between central and peripheral vision.

## Technical note

These results are taken from the 1987 ISO document ISO/TC159/SC4/WG2/N140. The original research papers were published by Hewlett Packard, Palo Alto, CA.

The graph shown in Figure 2-8 is presented in semi-log form. The horizontal axis—Display Refresh Frequency—is linear whereas the vertical axis—$E_{AC}$—is drawn to a log scale.

The semi-log presentation allows each Critical Flicker Fusion (CFF) Threshold curve to be displayed as a straight line. Each curve follows the empirical equation:

$$E_{CFF} = ae^{bf}$$

where the constants a and b were discovered from the experimental results which are set out in Table 2-1.

But the experimental results show something that is, perhaps, of even more fundamental importance.

There appears to be a constant 13 Hz overall perceptual difference in flicker between central and peripheral vision.

At low light levels, at 10 Trolands for example, the perceptual change in CFF, from central to peripheral vision, takes place between the 10 degree and the 50 degree contours. At medium and high light levels, in the range 1,000–100,000 Trolands, the perceptual change appears to take place much more swiftly between the 10 degree and the 30 degree CFF contours.

**Table 2-1**   Experimentally Determined CFF Constants
for Different Viewing Angles

| Viewing Angle | a | b |
| --- | --- | --- |
| 10 degrees | 1.705 | 0.0993 |
| 30 degrees | 1.162 | 0.0936 |
| 50 degrees | 0.301 | 0.1077 |
| 70 degrees | 0.594 | 0.0975 |

It is this snap change in flicker perception, occurring between 10 and 30 degrees, that probably underlies all the recent interest in VDT flicker.

## BRIGHTER SCREENS

Let us leave the TV viewer for a moment and return to the VDT operator. His normal working position is about 50 to 60 cm (20 to 24 inches) from the screen which has an active diagonal "working" area of about 25 to 30 cm (10 to 12 inches). A little trigonometry shows that the viewing angle of the operator usually ranges from about 28 to 34 degrees. The experimental results of the 30 degree CFF contour line which is drawn on the graph in Figure 2-8 is clearly now of some importance.

In the last 10 years there have been two major changes in how we use VDTs.

- The first change has come about from improvements in technology; the displays are now brighter than they were.
- The second change has come about from developments in software.

"Mouse Support," "Windows," "Icons" and "Pull-down Menus" have acquired levels of meaning that can never have entered the wildest dreams of the first lexicographer, Dr. Samuel Johnson.

The VDT user is no longer presented with bright letters on a dark background but rather the reverse situation now applies. We are now presented with darker letters set against a bright background, just like reading a very brightly lit book. It is this background screen brightness that exacerbates the flicker problem, especially when the information on the screen often directs our eye to the edges of the display rather than the center.

If, for example, we are concentrating on the left margin of the screen with our central (foveal) vision, this action allows a lot of light energy from the adjacent bright screen background to enter our eye. This light creeps in from the right of where we are devoting our attention and, rather unkindly, effectively stimulates the 30 and 50 degree peripheral CFF response contours of our eye. For the unenvied setter of VDT standards, it is thus the more demanding—all of 13 Hz more demanding—peripheral vision (30 and 50 degree) CFF contours that now rule the roost rather than our earlier ideas of central vision flicker susceptibility.

In setting these standards it is also important to take into account and to include the variation that occurs between different people in their perception of flicker. We will return to this inter-individual variation in CFF perception in a moment.

If the Troland inhabits the strange world at the back of the eye then it is the candela that inhabits the real world of bright television display screens. We have already traced the path from the bright TV screen with an illuminance of so many candelas per square meter (cdm$^{-2}$) through the pupil of the eye and finally into the Lilliputian world of retinal illumination energy, hammer drills and "thump" profiles.

We can now take a further step.

We can choose a range of values of typical screen illuminance and then check the pupil diameter that each screen illuminance will elicit. Once we know the pupil diameter we can derive the corresponding values of $E_{AC}$. We can then fit these $E_{AC}$ values to the experimental results which were presented a few pages earlier in Figure 2-8.

What we then obtain is our first direct correlation between screen illuminance and perceived flicker.

We can express this information in both a tabular and a graphical form. The tabular form is shown in Table 2-2, the graphical presentation is shown in Figure 2-9.

Although the title of the tabular results might appear cumbersome the caveat "95 percent of the population" neatly betrays the original purpose of the research. This was to establish a VDT display refresh standard that would satisfy nearly everyone—in this case 95 percent of us. Migrating the experimental results from tabular to graphical form proves much more revealing. The range of display screen illuminances appears as a shape not unlike a stepladder on the flicker threshold diagonals. The stepladder has deeper steps at the lower illuminances, relating to the wide spread in people's pupil size.

**Table 2-2**  Experimentally Derived Relationship Between
Screen Illuminance and the Refresh Frequency
Above Which Flicker Is Not Perceived by 95 Percent
of the Population.

| Screen Illuminance (cdm$^{-2}$) | Pupil Area (mm$^2$) | $E_{AC}$ (Trolands) | 10 deg CFF (Hz) | 70 deg CFF (Hz) |
|---|---|---|---|---|
| 10 | 10.75–19.63 | 215–392 | 48.4–54.4 | 60.6–66.9 |
| 30 | 9.08–15.21 | 544–912 | 57.6–62.8 | 70.2–75.5 |
| 100 | 8.04–11.34 | 1,608–2,268 | 68.9–71.9 | 81.8–84.9 |
| 300 | 7.07–9.62 | 4,242–5,772 | 78.2–81.3 | 91.4–94.6 |
| 1,000 | 6.16 | 12,320 | 88.9 | 102.4 |
| 3,000 | 4.52 | 27,120 | 96.7 | 110.5 |
| 10 k | 3.14 | 62,800 | 105.1 | 119.2 |

- At high display illuminances, 1,000 cdm$^{-2}$ and above, there is
  no significant inter-individual variation in pupil size and the
  depth of the step has become no more than a rung.
- At today's normal television set display illuminances, in the
  range 100–300 cdm$^{-2}$, there is only a 3 Hz inter-individual

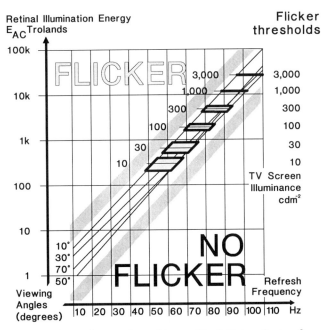

**Figure 2-9**  Experimental results—critical flicker fusion frequency plotted against screen illuminance.

difference in the perception of flicker which can be directly attributed to differences in pupil size at these light levels.

- For dim TV screens, in the range 10–30 cdm$^{-2}$, the difference in the perception of flicker rises to 6 Hz as we expect to encounter a two-to-one inter-individual variation in pupil area at these lower light levels.

It is worth recording that a doubling of pupil area leads to a downward 6 Hz frequency change in the onset of our perception of flicker. By the same token, a doubling in pupil diameter—which results of course in a quadrupling of pupil area—will cause about a 12 Hz downward frequency shift in our perception.

Although a doubling in the size of the pupil diameter might sound a little unexpected we do already experience this phenomenon every day. Do you remember how Figure 2-3 led us to expect a daytime pupil diameter of about 3 mm and a nighttime diameter of about 6 mm—an increase in size from an eighth to a quarter of an inch? We know that when we drive at night and catch sight of our neighbors' television set through a distant window, the neighbors' set seems to flicker far more than ours. We now know that this new-found sensitivity to our neighbors' television set is a result of a downward shift of about 12 Hz in our threshold of perception of flicker.

It can therefore no longer come as a surprise that Americans, accustomed to watching television with a 60 Hz refresh frequency, are appalled when they first see European television. The reduction in refresh rate of 10 Hz (from 60 Hz to 50 Hz) triggers a similar flicker response to what we experience when driving at night. It is perhaps capricious to suggest that, on our next visit to Europe, we might test the combined effects of a 10 and 12 Hz shift in perception by playing the two scenarios together. Catching a glimpse of distant televisions in a trip around UK residential areas at night should give the American visitor a good idea of what a 22 Hz downward shift does to our perception of flicker.

## THE STEPLADDER TEMPLATE

Now is the time to summon up our courage and, for a few moments, to suspend our well founded suspicion of statistics. We are about to discover that the "illuminance step-ladder" template, which we have just encountered in Figure 2-9, can be moved around the diagram with some interesting results.

But first we must make some simplifications.

We can reexpress the inter-individual variation in CFF frequency at each setting of the screen illuminance in terms of a mean and a deviation rather than as a range of values.

Thus, if we apply this technique to the experimental results in Table 2-2, we can express the mean peripheral vision CFF values of the dim 10 and 30 cdm$^{-2}$ screen illuminances as 51.4 and 60.2 Hz respectively and the total variability as ±3 Hz. At the more usual screen illuminances of 100 and 300 cdm$^{-2}$ the CFF mean values are 70.4 and 79.7 Hz respectively and the total variability is ±1.5 Hz. At high values of screen illuminance, 1,000 to 10,000 cdm$^{-2}$, the experimental evidence indicates that there is no significant inter-individual difference in the onset of flicker.

We may also infer from Table 2-2 that, in the three decade screen display illuminance range of 10–10,000 cdm$^{-2}$, there is a change in CFF frequency of 53.7 Hz. As the distance between each step of the stepladder can be seen to be quite regular (Figure 2-9), we can express the change in CFF frequency in a rule-of-thumb as follows:

        5.5 Hz    per doubling of screen illuminance

or

        8.5 Hz    per tripling of screen illuminance

or

        17.9 Hz    per decade increase of screen illuminance

or, better still, we can round the values off to 6, 9 and 18 Hz respectively.

Engineers have the same affinity to templates that more normal mortals have to jigsaw puzzles. Part of the engineer's training is to take a template or "pattern" and try to fit it on different parts of a graph just as we might try to fit a jigsaw piece into various parts of the jigsaw puzzle.

There is, however, a difference between the two techniques. The jigsaw piece is intended to fit properly only in one place whereas the template can provide different insights and understanding as and when it is moved from place to place.

Our stepladder template has two meaningful degrees of freedom:

- In the first instance it can be moved horizontally from side to side on the graph of CFF frequency versus screen illuminance.
- In the second instance it can be "slid" up and down the CFF diagonals.

## Inter-individual variability

We are all different. We all react differently to external stimuli, not least in our perception of flicker. Some people will perceive a picture as "steady" whereas other people will perceive the same picture as "flickery." Both judgments are valid.

The first example provided above—the horizontal movement—allows us to juggle the position of the stepladder template so as to fit each individual's response to flicker.

If I was watching a bright, 3,000 cdm$^{-2}$ screen then my own eye-brain might demand a screen refresh frequency of, say, 110 Hz to avoid flicker. On the other hand, my neighbor might be equally content with a screen refresh frequency of no more than 90 Hz in order to avoid flicker. The standard stepladder template can satisfy both our needs; the right hand stepladder in Figure 2-10 suits my own subjective eye-brain response whereas the identical left hand stepladder suits my neighbor.

Swedish research suggests that our perception of flicker exhibits the same pattern of statistical variability that we encounter when measuring many other natural events.

No one is quite sure why natural events follow the bell-shaped curve shown in the lower part of Figure 2-10 but it seems to work all the same.

Statisticians refer to this bell-shaped curve as a Gaussian distribution after Carl Friedrich Gauss who constructed useful sets of tables that allow us to make useful predictions from incomplete knowledge.

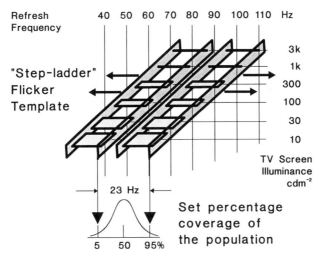

**Figure 2-10**  Template of CFF frequency versus display screen illuminance showing the percentage coverage of the
(Gaussian) population of viewers.

Opinion pollsters use similar techniques to gauge public
opinion by questioning a relatively small sample of people.

We are used to the idea of an "average value" and a "variation" about an average. The nomenclature of Gaussian Distribution statistics refers to the "mean" which is our "average value"
and the "standard deviation" (SD for short) which is a measure of
the variation in the results.

It may appear to be a contradiction in terms, but the Gaussian distribution is very precise about variability. This is perhaps
why it is so loved by mathematicians and physicists alike.

- It defines the range of one standard deviation either side of
  the mean (±1 SD) as encompassing 68 percent of all the
  variation we would ever expect about the mean value.
- Spreading our net a little wider, the range of two standard
  deviations (±2 SD) either side of the mean will encompass 95
  percent of the expected variation.

We can never quite catch 100 percent of the variation but we
can get as close to this limit as we choose.

Despite a faint whiff of magic about these results, the values of standard deviation are not just plucked out of the air but can be found relatively easily from published books of statistical tables.

One intermediate SD number that might not immediately spring to mind is the value 1.65 SD. This is the value that encompasses 90 percent of all expected variation about the mean. The other 10 percent escape our net, 5 percent are very much lower than the mean value and the other 5 percent are very much higher than the mean (i.e. more than 1.65 SD away from the mean).

- Swedish research indicates that, over the usual range of VDT display screen illuminances, the standard deviation for inter-individual differences at the 70 degree CFF contour is about 7 Hz. See Table 2-3.
- We also know that the American research results, shown earlier in Table 2-2 (page 127), offer VDT flicker limits that will satisfy 95 percent of the population.

From our knowledge of Gaussian statistics we can say that:

- The mean CFF frequency—where half the population will report flicker and half will not—lies 1.65 SD or 1.65 x 7 Hz = 11.5 Hz below the American "95 percent no flicker" experimental results of Table 2-2 and Figure 2-9.
- The lower 5 percent CFF frequency—where only 5 percent of the population will report no flicker—lies twice x 1.65 x 7 = 23 Hz away from the CFF frequency where 95 percent of the population report no flicker but a steady picture.

We are now able to move our stepladder template.

We may first move the stepladder template the horizontal "distance" of 23 Hz to the left—a distance of twice x 1.65 SD. This moves the template to the left from the 95 percent non-flicker figure to the 5 percent non-flicker figure. This is shown in Figure

**Table 2-3**   Inter-Individual Differences In Flicker Fusion Frequency.

| Mean screen luminance (cdm$^{-2}$) | 25 | 100 | 400 |
|---|---|---|---|
| Standard Deviation SD (Hz) | 5.71 | 5.78 | 8.29 |

2-10 as an operation which is intended to "set the percentage coverage of the population."

If we move the whole of the stepladder template only half this distance to the left, a distance of 11.5 Hz, then the template moves from the 95 percent no-flicker figure to the 50 percent figure. The template is now set at the top of the bell-shaped curve, at the mean value of CFF where one half of the population will claim to see flicker and the other half will not.

This is shown in Figure 2-11 below.

Figure 2-11 shows not one but two bell-shaped Gaussian distribution curves. The right one, which was originally demonstrated in Figure 2-10, represents the response of our peripheral vision, the new left one, set 13 Hz lower, represents the response of our central vision.

Let's try some numbers.

From the template we can see that, for a screen illuminance of 300 cdm$^{-2}$, just over two-thirds of the population, the 68 percent that lie within one standard deviation of the mean, will notice the onset of central vision flicker within ± 7 Hz of a 68 Hz display refresh rate.

Reducing the screen illuminance to 100 cdm$^{-2}$ will reduce the

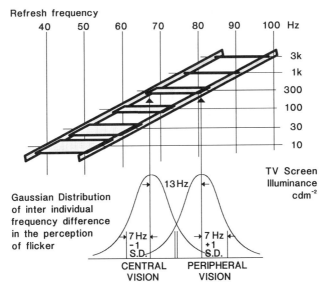

**Figure 2-11**   Template of population average CFF frequency versus display screen illuminance for central and peripheral vision.

onset of flicker by 8 Hz. The same two-thirds of the population will now notice central vision flicker within ± 7 Hz of a 60 Hz display refresh rate.

Do these results tally with our everyday experience?

Yes, I believe they do.

American 60 Hz TV is relatively flicker-free though it does show signs of flicker when the brightness control is turned up "too far." This inability to use the brightness and contrast of a modern TV set to its full extent is entirely due to the subjective limitation of our own flicker perception rather than a fault of the TV set itself. The 50 Hz Europeans, of course, have just had to live with the intrinsic flicker effect and this they have apparently done very well. It takes an American visitor to notice European flicker yet even this awareness wears off within days or weeks of his arrival.

Over this relatively long time span it would appear that our brain is able to "pull" the stepladder template some 10 or 15 Hz to the left as we adapt to the 50 Hz pictures. But, change the orientation slightly and the flicker comes back. Try this experiment for yourself. Clinical studies have shown that many of us will suddenly notice more flicker if the TV picture is scanned from bottom to top rather than from top to bottom. We can easily check whether we find this true by putting our head on one side and viewing the picture sideways on or up side down.

We do not yet know the price that we pay for this neurological adjustment to flicker. We do know, however, that visual display terminal operators can suffer headaches and possibly other subclinical symptoms if the display and/or the surrounding working conditions do not meet currently accepted standards. We also know that medical opinion suggests that using our brains to combat flicker makes as much sense as driving a car with the parking brake still applied.

**LOCKED IN**

This has been a tough chapter. We have not only explored medical science, the physiology of the eye and the application of

Gaussian statistics but have also discussed how a garden should best be watered and the inherent advantages of percussion drills for home improvements. What have we learned?

We have discovered a fundamental 13 Hz difference between our central and peripheral vision.

Why should this exist?

The 13 Hz difference probably dates back to pre-history, long before we even lived in caves. It is probably a manifestation of our defense reaction to DANGER, our intrinsic ability to spot a slight movement in our environment from the corner of our eye. Our peripheral vision is tuned to respond to movement—it registers flicker long after our central vision has decided that the image is steady, has no apparent movement and can therefore be safely discounted.

Once we detect a movement from the corner of our eye, our instinct makes us turn to look at that danger signal.

We find it hard to resist our inbuilt defense reaction that makes us turn our head to investigate the "danger" with our sharper central vision. We cannot help it, our attention will always be drawn to a flickery object.

The converse is also true. Our attention is not held by a steady or non-flickery object unless we have more pressing reasons for it to hold our interest.

With this in mind we can draw some interesting comparisons between today's conventional and high definition television.

Normal television sets, 50 Hz European or 60 Hz American, exhibit flicker. Our attention is, quite naturally, drawn to them when we enter a room because we first see the screen from the corner of our eye.

However, when we then look at the screen directly (with our central vision) the flicker mostly disappears.

The current proposals for HDTV receivers accept that flicker must be eliminated from the high definition pictures. Some HDTV sets use the relatively simple technique of "frame doubling" which raises the field rate of the picture from 50 to 100 Hz in Europe and from 60 to 120 Hz in America.

This technique works only too well—the flicker is gone.

But have we thrown out the baby with the bathwater?

If we look at Figure 2-11 we can see that, at a typical screen illuminance of 300 cdm$^{-2}$ and a refresh frequency of 100 or 120 Hz, all trace of flicker has been banished from both our central and peripheral vision.

But this is not the good news it might at first appear. We no longer have any pressing reason to look at the screen.

Our peripheral vision no longer shrieks DANGER to our brain or makes us turn our head—involuntarily—to see from whence the danger might come.

The elimination of flicker has, in effect, destroyed our interest in looking at the screen.

It is now no more interesting than a brightly lit picture hanging on the wall. No wonder that many HDTV researchers have complained that the HDTV pictures might be sharp but also appear to be "dead" and "lifeless" when compared to our familiar flickery conventional television pictures.

## A REFRESHING STRADDLE

To regain the viewers' attention we must re-introduce some peripheral flicker.
Figure 2-12 indicates how this can be done.

At the normal screen brightness of 300 cdm$^{-2}$, most of us will normally first discern the disappearance of central vision flicker at 68 Hz and the loss of peripheral flicker at 81 Hz (give or take a few Hz).

We can now play a trick on our flicker perception.
Imagine a normal brightness television screen which is refreshed not at 50 or 60 Hz but at 75 Hz. You will notice that the refresh frequency of the display has been set above the 68 Hz threshold at which we no longer discern central vision flicker.

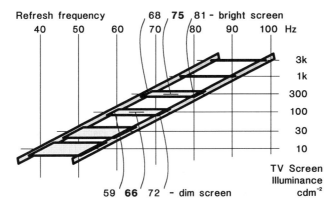

**Figure 2-12** Template of population average CFF frequency versus display screen illuminance showing how the central—peripheral vision CFF frequencies may straddle the display refresh frequency.

Thus, when we look straight at the television screen, it appears quite steady—the flicker has vanished.

However, the 75 Hz screen refresh frequency is still not sufficiently high for our peripheral vision also to report the loss of flicker. Figure 2-12 suggests that we would need to raise the refresh rate to at least 81 Hz if we are also to pass that threshold.

When we look at the screen indirectly, from the corner of our eye, the screen still appears to flicker.

"DANGER—look at the screen!" shrieks our cave man defense instinct.

But, when we turn to look, the picture is rock steady, easy on the eye and easy to watch—unlike the conventional 50 Hz PAL and 60 Hz NTSC television pictures which have permanent residual, discomforting central vision flicker unless the brightness is turned down. Often enough our TV set will perform this task for us by automatically turning down both contrast and brightness when the room lighting gets dim.

If our attention should wander and we should dare look away from the screen, the cave man danger cycle will repeat itself, our head will turn and the screen will capture our attention once more.

- We are locked in to the TV program and, more to the point, are locked in to the advertisements that follow.
- At our deepest biological response level the picture is simply irresistible!

In more scientific terms we can say that we have locked in the viewer by setting the screen refresh frequency midway between the central and peripheral flicker thresholds anywhere on the template.

Figure 2-12 allows us to determine that a relatively dim screen, less than 100 cdm$^{-2}$, will lock in the viewer at a refresh frequency of about 65 Hz whereas a bright screen, 1,000 cdm$^{-2}$, would exhibit lock-in at a refresh frequency of 85 Hz.

This is of only academic interest, however, because we cannot expect a national TV system to change its picture refresh rate from 65 through 75 to 85 Hz just to suit the convenience of how bright we might each set our own TV picture.

But the idea is not entirely lost.

The choice of 75 Hz as the picture refresh frequency is based on the brightness of a typical domestic television set with a conventional cathode ray tube as the display element.

On such a domestic TV screen the 75 Hz refresh frequency is ideal but, if we choose a different display device, such as a liquid crystal display panel (an LCD), then the flicker parameters are quite different. In technical terms the $E_{DC}$ conditions remain the same but Amp$(f)$ (the thump factor) and hence the $E_{AC}$ parameter, will change significantly. Whereas the CRT thump factor stuck stubbornly to a factor of 2, the LCD can be set over a range of values but with "2" as its upper limit. How the Amp$(f)$ thump factor parameter can be easily changed in practice is the subject of a European patent granted to the author.

There are two application areas.

**Commercial display**

Outdoor television displays need to be quite bright if they are not to appear "washed out" when the sun shines. If the high brightness is obtained from conventional cathode ray tubes, then a screen refresh of 100 Hz or more is needed to avoid flicker.

However, if we can modify the thump factor, the value of Amp($f$), by the use of LCD elements in place of the CRTs, then we may reduce the refresh frequency from about 100 Hz to 75 Hz and yet still keep the display subjectively free of flicker. With care we can further fine tune the thump factor so as to set the screen flicker effect near the middle of the 13 Hz gap which exists between our central and peripheral vision. At this point we not only provide a flicker-free picture, we can also lock the viewer into watching the display.

### Domestic display

If the future domestic television set is to abandon the cathode ray tube (with its fixed thump factor) then we might enjoy the "75 Hz lock-in effect" on, for example, an LCD screen which is refreshed at the more familiar frequencies of 60 and perhaps 50 Hz.

We can see from Figure 2-12 that, in order to "pull" our flicker perception threshold 20 Hz downwards from, say 100 to 80 Hz, then the effective screen flicker brightness ($E_{AC}$) must be reduced tenfold, from 3,000 to 300 cdm$^{-2}$.

Alternatively we could reduce the thump effect from a value of 2 to 0.2.

In practice, it may be unrealistic to "pull" the flicker threshold more than about 20–25 Hz, though, as we discovered earlier, the European brain has learned to accommodate an additional downward "pull" of 10–15 Hz in long-term flicker perception.

It would appear sensible to include the perception of flicker in the parameters that will define a viable high definition television standard. It is clear that we can enjoy a number of advantages if we choose a picture refresh frequency of somewhere near 75 Hz.

- Such a refresh frequency is ideal for a CRT based domestic television set which can now be operated at higher brightness than today's relatively dim HDTV 50 or 60 Hz receivers.

- It can also be used without difficulty in an LCD based domestic television set.
- Such a no-flicker refresh rate is within reach of a high brightness outdoor television display provided that it employs LCD elements or lamps rather than the traditional CRT as the display element.

### Subliminal television

It is often claimed by the practitioners of the art that the layout of the merchandise in a large store has been scientifically planned, with promotional items at the end of lanes, high-profit margin items set on shelves in the wide lanes and regular lower-profit items tucked away in narrower lanes at the back of the store.

The typical customer is said to progress around the store in a predictable way, though I am not sure whether this perambulation is intended to include the search for a bag of sugar which has been moved unexpectedly four rows back.

Might we apply this same technique to the everyday 30 second television advertisement by carefully directing the viewer's attention first to one part of the screen and then to another?

> By dynamically modifying the thump factor in just a part of the picture we can arrange to draw attention to the merchandise or the presenter in turn.

The dirty towels that were washed in brand X might be displayed at one flicker parameter whereas the results from the adjacent new improved Brand A product can be displayed at a flicker setting that makes us feel more comfortable.

We can apply exactly the same technique to the televised version of the feature film.

The "good" Captain Kirk can be shown as a rock steady flicker-free character whereas the wicked Klingons can be shown in a "bad light."

The horror story reaches its dreadful climax, the tone of the background music grows ever more menacing and the picture begins to flicker disconcertingly.

When will "Jaws" re-appear for the two note theme music has started to play quietly once again?

Our eyes are drawn first to one part of the scene then another.

Some scenes can be flicker free and re-assuring.

We are safe: Jaws is not there.

We suddenly notice that the next scene has a hint of flicker in the corner.

Is that where Jaws is hiding?

Will we be scared to death?

Yes.

Will we trade in our new high definition television set for the regular model which does not have scary pictures?

No way!

Dynamic flicker HDTV is to regular TV what color TV was to black and white. It is the difference between FM stereo and AM Mono.

Dynamic flicker HDTV is not just a better version of the regular mousetrap—it is a new viewing experience and we could use it NOW. It opens up a new market for LCD TV sets as the full effect can only be created on the display screen itself.

How we might employ this effect in practice is discussed in the next chapter entitled "Simply Irresistible."

# 3

# Simply Irresistible

## INTRODUCTION

Back in the balmy spring days of early 1987 it seemed that there were few clouds in the sky. The world economy was climbing steadily and Vice President Bush was set to win a 4 year term in office. The Black Monday stock exchange crash of October 1987 and the fall of the old guard in the Soviet Union were not even distant specks on the horizon.

Around the world HDTV was making slow but steady progress. In Japan the State broadcaster NHK had spent 17 years and vast sums of money on research and development. The 1,125 line Japanese HDTV cameras and monitors on show in America were able to send sparkling pictures across the studio. HDTV satellite transmissions were planned for the future.

Europe was also keen on HDTV. In technical terms the HD-MAC TV transmission system was very similar to the NHK MUSE system though the Europeans claimed theirs was an evolutionary not a revolutionary approach.

What of the American developments?

A lot of work had been accomplished. The main thrust of American research had been in widening and/or sharpening the existing NTSC color picture. Ives Faroudja had developed a number of smart boxes that could fool the eye into believing the NTSC picture was sharper than it really was.

At MIT Russell Neuman and William Shreiber were also winning industry funds. Schreiber was concerned about the effect of new TV developments on US jobs and regretted the US government's meek acceptance of the foreign 1,125 line standard. Neuman begged us to look beyond brighter colors and wider screens to an open public discussion of quality versus quantity. He proposed an "Open Video Architecture" that could embrace telephone, computer and broadcast expertise so as to provide the greatest range of programming at the lowest possible cost.

For seven years William Glenn at the New York Institute of Technology had been working on an NTSC based HDTV system which would use an augmentation channel to sharpen the basic picture. At the David Sarnoff Research Center James Carnes' team proposed in-channel enhancements to NTSC that could be later upgraded to full HDTV when additional bandwidth became available.

And then the thunderbolt stuck.

The American Land Mobile Communications Council (LMCC) requested the FCC for more frequencies for cellular telephony. They had spotted that, of the 55 UHF channels available for broadcast TV in the United States, technical considerations dictated that no more than nine would ever be in use in any locality. The cellular business was booming, major markets were becoming congested and the LMCC were intent on putting these idle channels to more profitable use.

The broadcasters were appalled. Sure, there were some spare frequencies but these gap channels were needed to protect the existing broadcasters from mutual interference.

Or were they?

Cable systems have no gaps, for cable can only make a profit if 30 or more channels are stacked up next to one another. But this gap-less technology does not come cheap. Cable systems use very high-quality specialist components and need good technical help if their networks are to be set up properly.

The LMCC argument for reassigning frequencies seemed overwhelming. Would the broadcasters be able to defend their long-held frequency spectrum? An aggressive public relations campaign and a program of technical innovation were chosen as the two lines of defense.

Powerful broadcast lobby groups swung into action. Did the American public really want selfish high price business telephone users to triumph over a universally available over-the-air family entertainment television service? The National Association of Broadcasters (NAB) even boasted that "competition has driven the US broadcasting industry to produce the finest and most sophisticated broadcast service in the world today!"

The technical arguments were more measured.

Yes, there were a lot of frequencies left unused but this was directly related to the cost of the regular TV set. If less frequencies were to lie fallow then the cost of TV receivers would have to rise by $10–25, the same level of capital cost as a set-top cable hookup. Such improvements could not be introduced merely at the whim of technical idealists—there must also be a noticeable benefit for the TV viewer.

But what should the benefits be?

There was certainly no shortage of good ideas to be found in the many research labs around the USA. How could they best be harnessed?

## ADVANCED TELEVISION IN THE UNITED STATES

On February 13, 1987, 58 interested parties jointly approached the American Federal Communications Commission (FCC) with a "Petition for Notice of Enquiry." They argued that the Commission should consider the future of Advanced Television (ATV) prior to allowing any further sharing of the UHF broadcast spectrum by the Private Land Mobile Radio Services.

Six months later, on August 20, 1987, the Federal Communications Commission issued a Notice of Inquiry (NOI) in the Matter of: "Advanced Television Systems and their Impact on the Existing Television Broadcast Service."

As is customary in the United States the FCC invited public

comment not only on the likely course of development and the value of terrestrial advanced television but also on its probable impact on existing TV broadcasts. The public response was a fascinating mix of views. It highlighted different facets of the challenge that would face any proposed improvement in television service anywhere in the world. Many of the technical issues remain as fresh today as when they were first written.

The submission from Capital Cities/ABC put the issue in a nutshell—

> The aim is not to achieve technical perfection but to provide a service that can compete successfully for the patronage of the general public.

Compete successfully of course against cable, satellite and video rental.

At first the cable TV industry sat on the fence. Sure they would support ATV but the new signals must be designed to be "cable friendly" as well as being easy to transmit over the air. However, there was a sting in the tail. If the efforts to make the signals "cable friendly" meant that the broadcast pictures were not as sharp as they might otherwise have been then the cable industry reserved the right to develop their own ATV system—in order to keep up with the growing competition from videocassette and laserdisc. It is reassuring to know that since that time the cable and broadcast industries have recognized their common interest and have worked closely together on developing ATV.

The satellite industry took the view that the terrestrial frequency spectrum issues were so complex that they would take ten or twenty years to resolve. Not surprisingly they suggested that we should leave the terrestrial mess behind us and start again. Backyard receiving dishes could pick up wider bandwidths and hence produce sharper pictures.

For the last ten years the idea of abandoning terrestrial broadcasts in favor of satellite transmission has been a familiar theme throughout the world. In the early 1980s the Europeans developed MAC for satellite-based television distribution. The Japanese have taken the same view for their MUSE HDTV system.

What drove sensible engineers to invest $200 million a time in launching high-power direct-to-home TV broadcasting satellites? It was certainly not a love of gambling for their hearts were in their mouths every time they watched a new satellite lift off from Cape Canaveral. Theirs was a different fear, the fear that the terrestrial spectrum had become too complex for them to control.

It is the same fear that can overtake us when we try to upgrade the software in our personal computer. The instructions seem straightforward enough. Add three files here, reset the PIF parameters there, refer to the small print in the manual if the DOS software revision level is below 3.3. All so easy for budding Einsteins. These days I cheat. I delete the whole application and start again from scratch. Despite an earnest wish to repack the terrestrial spectrum the frequency planners cannot enjoy such luxury—they have no choice but to work with what has gone before.

## NTSC complications

When black and white television was first introduced over fifty years ago the transmission consisted of two radiofrequency signals—the major (video) picture carrier and the additional minor sound subcarrier. In 1954 the NTSC color system added another signal, the color subcarrier, which was placed in a convenient gap between the two existing signals. More recently stereo sound has been introduced by the addition of yet another subcarrier next to the original mono sound signal. Today nearly every American NTSC TV channel is a carefully crafted mix of four separate signals.

Why does this cause so much difficulty?

Imagine driving down a multilane highway on a rainy night. All the traffic coming towards us has its lights switched on. Let's say that the two white headlamps on each automobile represent the TV video and sound signals and that the two amber side lights represent the weaker color and stereo sound subcarriers.

But it is raining hard and the windshield wiper blades are long past their best. We see multiple images and false reflections through the traffic grime on the windshield. Is that a white headlamp in the outside lane or is it a reflection from an adjacent lane? Does that amber light belong to the second or the third vehicle? Darn it, there is too much oncoming traffic on the highway to make out what is what.

If we are the driver of the sedan then the cable TV operator must be the trucker. He is able to sit up high and gets a better view of the road ahead. He changes his windshield wiper blades regularly thereby avoiding many of the spurious reflections that we experience. He can also filter out some of the oncoming lights by pulling his sun visor right down low or by moving slightly back and forth in his seat. Unlike us he does not mind solid lines of traffic in every lane on the highway.

For all his experience and professionalism, however, the truck driver does have a weak spot. He can deal with four lights on a vehicle but the addition of one or two extra spot beams can cause him utter confusion. Thus the cable TV operator is entirely averse to schemes that add any more subcarriers to the conventional NTSC signal. Taken thirty at a time even "simple" NTSC signals can cause trouble. Adding further complexity to each transmission makes the cable operators' task far too difficult for widespread adoption.

What of our European cousins? How do they fare under similar driving conditions?

United Kingdom viewers currently enjoy a national four channel terrestrial television service though there are plans to introduce a fifth channel in a few years' time. Unlike the position in the USA it is customary for UK viewers to obtain television service from a nearby transmitter mast on which all four TV channels are co-located. It has thus proved relatively easy to encourage city dwellers to install a single high performance rooftop antenna through which they can receive all terrestrial television programs. As a result the quality of domestic pictures is extremely good and less than 3 percent of UK households subscribe to cable.

Unlike the VHF FM radio broadcasting service, where the use of low performance telescopic whip antennas is almost universal, there is a strongly ingrained public perception within the UK that good television reception can only be obtained through the use of a properly installed rooftop antenna. This perception is

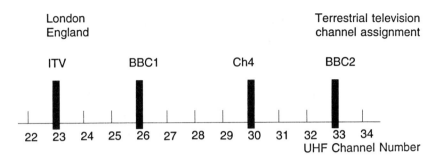

**Figure 3-1** The London taboos. In the London area those channels
either side of ITV, BBC1, Channel 4 and BBC2 are left
deliberately unused. Many of them can be assigned for
use by low-power digital television without interference to the existing services.

perhaps self-fulfilling. UK television transmitter powers (and
hence electricity costs) are significantly less than would be needed
if it was generally thought that set top "rabbit ears" antennas
were able to give good pictures. Whereas US TV transmitters can
often exceed 5 MegaWatts (MW) in power most UK transmitters
rarely exceed a tenth of this figure.

The UK frequency planners have been able to design a UHF
channel allocation system which currently provides four high
quality pictures to more than 99.4 percent of the population. The
London area channel assignments, Channels 23, 26, 30 and 33
are typical and are shown in Figure 3-1.

### It's taboo

Perhaps the most striking feature of this assignment pattern
is the large number of channels left unused.

Let us return to our highway analogy for a moment. Imagine
an urban highway which can carry four lanes of traffic—two lanes
in one direction and two in the other. In the UK such a highway
is normally described as a two lane dual carriageway. When the
road is first planned each lane is assigned a standard width of
14 feet with the result that each carriageway, containing two
lanes, is 28 feet wide. Experience has shown that it is too dan-
gerous to place the two 28 ft. carriageways immediately side by
side. Instead they are separated by a no-man's-land or central
reservation that is itself 6 feet or more in width. When we include

two grass shoulders in the calculation the total width of the urban highway is not far short of 80 feet.

Is this reasonable, an 80-foot highway for four cars, each no more than 6 feet wide?

Yes, it is reasonable. We gladly acknowledge and accept that the average motorist needs his share of "elbow room" on the road.

We can now read and interpret Figure 3-1 in a new light.

- The "outside" channels, 22 and 34, are the grass shoulders.
- The middle channel, Channel 28, is the central reservation.
- Channels 24, 25, 27, 29, 31 and 32 are the essential gaps, the motorists' "elbow room," that is needed between the four principal channel assignments.

Throughout the world the frequency planners have adopted similar "elbow room" schemes though there are considerable variations in approach. Some countries, such as the UK and Japan, have four or six national terrestrial channels in which a relatively small number of high-power transmitters can reach 70 percent of the population. The remaining 30 percent are served by a large network of medium and low-power repeaters which each need just as much care in planning their frequency assignments.

The US approach is quite different. Because of the commitment to "localism" there are larger numbers of primary transmitters and relatively low numbers of repeater stations. Just compare the figures. The UK, which is a little smaller than either Oregon, Michigan or Wyoming, has 200 high-power transmitters and nearly 4,000 repeaters that serve 23 million homes. The USA has about 1,700 primary transmitters and 6,500 low-power and/or repeater transmitters to serve 92 million homes.

Let's recap.

The cable TV engineers are opposed to the addition of any more subcarriers in existing NTSC channels—it is tough enough to get a good cable performance as it is.

The frequency planners have good reason to keep many of the channels unused, so as to provide "elbow room" for the main channel assignments.

No wonder everyone turned to satellite transmission as a way out of these constraints.

But the door has not quite closed in our faces. Let's try a little lateral thinking.

We may not be able to add a totally new subcarrier to the NTSC signal but we may be able to make some substitutions. For example if we choose a letter-box format for the picture we could place data signals in the black bands which now appear at the top and bottom of the picture. If we use low-power PSK, the workhorse of digital communications, perhaps we can craft its introduction so that we will hardly notice the signal at all. We first met this technique in Chapter One in its European form, PALplus. The Japanese have made similar proposals for improving their regular NTSC service.

And now for the real lateral thought.

If we can risk placing low-power digital PSK signals in the on-screen black bands then might we not also broadcast the same PSK signals in the adjacent "elbow room" channels that are better known to American broadcasters as the taboo channels?

Let's revisit the highway analogy. When we are stuck in traffic the four-wheeled motorist is often irritated by the more nimble two-wheeled motorcyclist who is able to weave in and out of the lanes of automobiles in front of him. Surely such blatant lane changing is taboo! The local police department probably think so. On the other hand the more mature motorcyclist, a rare breed, would argue that no purpose is served in standing still in a line of traffic when his small size permits him to make headway without any disadvantage to the line of stalled automobiles.

Can we observe the same rules of co-existence in our use of the taboo television channels? On the one hand the low-power digital PSK taboo channel transmission, the motorcyclist, might claim the right to go anywhere but, in so doing, would disadvantage the existing high-power television NTSC broadcast, the motorist. On the other hand the motorist cannot disagree that, in a tightly regulated environment, when the traffic is flowing slowly there is ample room for the motorcyclist to drive down the white lines that separate the lines of cars.

## Digital Augmentation

We are still in with a chance. It looks feasible to add a digital augmentation channel alongside an existing NTSC channel in order to carry "picture sharpening" information.

Where is the augmentation channel to be placed?

Should it be next to the existing NTSC channel or could it be somewhere else on the dial. This is the same problem that we experience when booking airplane tickets. Getting two seats on the plane is usually not that hard but getting two seat assignments next to one another takes much more work, especially when the airplane is nearly full.

The TV set-makers were not slow to point out that the exact placing of the augmentation channel will have a marked effect on the cost of implementation. If the augmentation channel is not placed alongside the main NTSC channel then the TV set will need two tuners, one for the main NTSC channel and another completely separate one for the augmentation channel somewhere else on the band. If luck is against us there may not even be a spare channel available in the same band. Imagine the complication if the existing main NTSC channel is on VHF and the augmentation channel can only be fitted in on UHF. Could viewers be persuaded to go to the expense of a UHF antenna just to sharpen the VHF NTSC picture?

Was there a middle course?

Experiments had shown that under laboratory conditions NTSC pictures could be both widened and sharpened by the addition of a modest amount, about 1 Mb/s, of digital augmentation. Calculations showed that such a signal did not need to use the full 6 MHz of a taboo channel, in fact two or three 1 Mb/s augmentation signals could be made to fit into one taboo channel quite nicely.

Looked at another way the standard 6 MHz NTSC was being widened to 9 MHz by the addition of 3 MHz of augmentation bandwidth. Quite often a new 3 MHz slot could be found either just below the station's existing NTSC transmission or sometimes just above. This was not entirely convenient as the two options required a slightly different technical implementation. Less often a new allocation could only be found above or below a competitor's station frequency. In this case it was unclear from which tower the augmentation channel should be broadcast. Preliminary studies showed that in the important and hence congested major markets there were few opportunities to increase a station' bandwidth from 6 to 9 MHz and so the idea was eventually dropped.

The Japanese were disappointed. Their 1,125 line MUSE HDTV system had been engineered to work in the satellite equivalent of an 8.1 MHz channel so a 9 MHz terrestrial channel would have suited their system very well. Reducing the bandwidth to 6 MHz had an adverse effect on the quality of the Japanese picture.

The flirtation with non-standard channels was over.

Whatever picture improvements might be offered to the terrestrial TV viewer in years to come it was clear that they would have to be delivered in a 6 MHz channel.

Only two choices were left.

1. Adding hidden signals to a full-power NTSC channel.
2. Broadcasting low-power signals in taboo channels away from the main NTSC channel.

The broadcasters knew all about the first option for they had taken considerable flak from cable operators when they had converted their stations to stereo sound. Adding still more signals within the NTSC channel might still be a possibility but it would be an uphill struggle.

The second option began to look more attractive. The broadcasters already knew that a 1 Mb/s augmentation signal was sufficient not only to sharpen an NTSC picture but to construct the additional sidepanels of the wide-screen picture from scratch. Why not dispense with the NTSC picture all together and make the transmission entirely digital?

The race was on.

On one side of the fence the picture compression experts were producing sharper and sharper pictures from lower and lower digital data rates. On the other side of the fence the broadcast engineers were persuading Semaphore man to wave his arms ever faster.

## THE AMERICAN CLASS OF 1991

Following the frenetic activity in the summer of 1987 the FCC was very soon overwhelmed by many rival claims for high definition and improved definition television systems. A year later it issued a "Tentative Decision and Further Notice of Inquiry" in which it suggested that attention should first be focused on improvements to VHF/UHF terrestrial broadcasting rather than the wider issues of satellite, videocassette and laserdisc standardization. In March 1990 the FCC announced that a television standards decision was planned for early 1993 and that this decision would be based on the practical results of a competition to be held between rival systems. Testing was conducted in cooperation with three laboratories, The Advanced Television Test Center (ATTC) and the Cable Television Laboratories (Cable-Labs) in the United States together with the Advanced Television Evaluation Laboratory (ATEL) in Canada.

On April 15, 1991, Richard E. Wiley, the Chairman of the FCC Advisory Committee on Advanced Television Service, addressed a meeting of MSTV (Maximum Service Television) in Las Vegas with the following words:

Just a few years ago, when the Advisory Committee first got started, the conventional wisdom about advanced television service was that:

1.  The transmission mode would be analog,
2.  The United States was far behind other nations of the world, and
3.  Broadcasters would not be a real factor in this advanced environment. But, today, here are the facts:
    a.  Four of the six proponents now present simulcast HDTV system concepts, employing an all-digital transmission format. This has required the Advisory Committee to make changes in our test program.
    b.  In many respects the United States is now a leading force in advanced television system development. The decision that the FCC plans to make in mid-1993 to establish a new terrestrial broadcast standard—which may involve a digital format—could set

the pace for the rest of the world relative to this manner of introducing HDTV.

c. And finally, it is clear that broadcasting and cable are in the immediate forefront of the FCC's plans for HDTV. The FCC believes that terrestrial transmission will be the quickest and the least expensive way to introduce this exciting new service to the American people.

Without question, NTSC will remain an important element of television broadcasting for some years to come but it also will be forever a narrow-screen medium and inherently restricted by the constraints of the current standard. In contrast, advanced television service represents an exciting new and limitless video frontier.

Clearly, this new service can be delivered via cable, DBS and, ultimately, fiber optics—that is broadband transmission media.

But terrestrial HDTV is now also potentially possible, due to some exciting video compression breakthroughs that have been achieved in the last year or so.

This development may make it possible for your industry to emulate, within the American home, the exciting dimensions, sound and clarity of the motion picture viewing experience. And it also could give broadcasting, for the first time, access to the multi-channel world that your competitors enjoy.

I would respectfully suggest that broadcasters cannot afford to be second best in this dynamic environment.

## The 1991 proponents

The proponents were drawn from four groups who, between them, offered six rival ATV systems.

### The four groups

1. **American TeleVision Alliance (ATVA)**
   Partners
   General Instruments Corp.
   Massachusetts Institute of Technology (MIT)

2. **Advanced Television Research Consortium (ATRC)**
   Partners
   National Broadcasting Corp.
   North American Philips
   David Sarnoff Research Center
   Thomson Consumer Electronic Inc.
3. **Digital Spectrum Compatible (DSC)**
   Partners
   Zenith
   AT&T
4. **Japan Broadcasting Co (NHK)**

### The six systems

All six systems were designed to operate in a standard 6 MHz channel. The analog systems were intended to operate at conventional high power whereas the digital systems operate at the low power levels consistent with the use of the taboo channels.

### Two analog ATV systems

1. **Advanced Compatible Television (ACTV).** The ATRC consortium bid ACTV. Of the six proponents it was the only truly compatible NTSC-EDTV system as it enhanced a conventional NTSC picture. ACTV uses a number of in-band coding techniques to increase the horizontal definition from 330 to 700 lines and to widen the aspect ratio. These techniques had been developed over a number of years by Dr. James Carnes' research team at the David Sarnoff Research Center.
2. **Narrow MUSE (NHK).** Standard MUSE was designed to work in a 9 MHz bandwidth which is readily available on satellite. In order to fit the MUSE signal within a standard 6 MHz terrestrial channel NHK therefore proposed two lower quality formats, NTSC MUSE 6 and Narrow MUSE. Just like the ACTV pro-

posal described above NTSC MUSE 6 was compatible with existing NTSC sets but did not produce such sharp pictures as Narrow MUSE.

Narrow MUSE used a transmission method similar to the time compression technique used by European MAC and was therefore not directly compatible with existing NTSC television sets. It achieved the bandwidth reduction from 9 to 6 MHz by first down-converting the studio HDTV signal from 1,125 to 750 lines (1,035 to 690 visible lines). At the receiver the proposed Narrow MUSE domestic decoder up-converts the picture for display on the *de facto* "standard" 1,125 line interlaced screen. We shall discuss these down and up conversion techniques in more detail further on.

In the event the analog ACTV system was later withdrawn in favor of the consortium's alternative proposal—Digital AD-HDTV. Narrow MUSE was thus the only analog system to be considered.

### Four digital ATV systems

The four digital transmission television systems are expressed in Table 3-1, as follows:

AD-HDTV—Advanced Digital High Definition Television was bid by the ATRC group—NBC, Philips, Sarnoff, Thomson and Compression Labs.
Digicipher—(interlaced version) was bid by GI and MIT
DSC—Digital Spectrum Compatible was bid by Zenith and AT&T
CCDC—Channel Compatible Digicipher (pro-scan version) was bid by GI and MIT.

Perhaps the most remarkable feature of the four digital proponent systems is their similarity. The ATRC consortium bid a 1,050 line interlaced system yet its operation is heavily dependant on internal conversions to and from progressive scanning.

The General Instrument—MIT "Digicipher" collaboration takes us one step further for they hedged their bets by bidding

**Table 3-1**    Overview of the Four Digitally Based Proponent
Advanced Television Systems.

| Type | No. of Lines | Pixels h & v | Scan | Video Mb/s | Gross Mb/s |
|------|------|------|------|------|------|
| AD-HDTV | 1,050 | 1,500h −960v | inter | 17.7 | 24.0 |
| Digicipher | 1,050 | 1,480h −960v | inter | 17.47 | 24.39 |
| DSC | 787 | 1,280h −720v | prog | 16.92 | 21.0 |
| CCDC | 787 | 1,280h −720v | prog | 18.88 | 26.43 |

both 1,050 interlaced and 787½ pro-scan. Zenith and AT&T had little doubt about the right way to go. Their DSC bid was entirely pro-scan.

There is also a considerable amount of agreement in the choice of the video digital data rate, varying from 17 to 19 Mb/s. The gross data rate is significantly higher than the net video data rate for two reasons.

1.  It includes audio channels, closed captioning and other house-keeping information.
2.  It includes additional "packing material" which is intended to protect the net signal against transmission errors. More of this later.

It appears that the standard terrestrial 6 MHz channel allocation will need to support a gross data rate of about 21 to 27 Mb/s and the proponents suggest a number of ways in which this data rate can be successfully squeezed into the 6 MHz television channel.

The first indications from the 1991–1992 test cycle show that the taboo channel digital transmissions—our motorcyclists who drive down the white lines—will need beefing up if they are not to be buffeted too strongly by the slip stream from existing NTSC broadcasters.

Computer predictions indicated that, before optimization, only 30–35 percent of digital transmissions can promise low-power transmissions of 100 kW or less. It is expected, however, that refinements to the planning model will raise this percentage considerably. All the digital ATV proposed systems claimed to provide a service area very similar to existing NTSC transmissions.

By comparison the same process indicated that only 5 percent of Narrow MUSE transmissions could operate successfully at 100 kW or less. As a result the computer program was able to find suitable frequencies for only 1,582 of the 1,657 required assignments and the service area of many of these allocations is considerably less than NTSC. These range limitations of the Narrow MUSE system soon led US industry to concentrate on all-digital ATV systems.

Many telecommunications engineers believe that the over-the-air data link to our homes will prove more of a technical challenge than the design of the video compression algorithms which are at the heart of digital HDTV. Although the proposed data transmission techniques are not new they have never before been widely implemented at such high speeds.

- On the one hand the Europeans already have experience of transmitting 20 Mb/s data over the air. This is the data portion of the hybrid D-MAC signal but takes about 12 MHz of bandwidth rather than the 6 MHz on offer.
- On the other hand for many years we have been able to transmit high-speed data over low bandwidth dial up telephone connections by the use of computer "modems." These boxes perform the same semaphore-like manipulations on the data, albeit much slower, that the ATV proponents suggest for squeezing 25 Mb/s digital television into the 6 MHz terrestrial TV channel.

Recent work on Terrestrial Digital Audio Broadcasting T-DAB has opened up another line of development.

## OLYMPIC SEMAPHORE MAN

In the summer of 1992 it was Spain's turn to host the Olympic Games. On the first day the stadium was filled to overflowing as many famous Spanish singers and dancers took part in the lavish opening ceremony. Although a few took along their own binoculars in order to follow the action in more detail most spectators chose to watch the giant television screens as Jose Carreras sang arias from well-loved operas.

For part of the ceremony, however, the high-power lenses could be put away for the audience in one part of the stadium had been trained in the use of brightly colored flags. On a given signal they waved and changed these flags in unison so as to create a series of colored pictures that moved, merged, wiped and cut from one scene to another.

This multiple flag waving differed from the individual artists presentations in three ways.

1. From across the stadium it was difficult to pick out individual flags with the naked eye yet the overall effect was unmistakable.
2. Although each flag waver changed the position of his flag relatively infrequently it was important that the change occurred at the right moment.
3. Each flag waver had to synchronize his arm movements with his neighbors in order to avoid unexpected clashes.

The European Broadcasting Union has developed a new form of data transmission which is based on these principles. It is called Coded Orthogonal Frequency Division Multiplexing—COFDM. "Coded Orthogonal" means that the flag wavers take measures to avoid hitting one another. "Frequency Division Multiplexing" means that each of the many flag wavers has his own job to do.

In this scheme one high-power fast-moving semaphore man is replaced by hundreds or thousands of smaller semaphore men, each waving their flags more slowly.

COFDM has two advantages over standard semaphore.

1. It can operate at even lower power than standard semaphore. We need high-power binoculars to see Carreras sing whereas the crowd of Olympian flag wavers can be seen clearly with the naked eye.
2. Each individual's flag waving speed is so slow that a schooldays' "copycat" can mimic every flag movement

without error. This is a tremendous advantage. In technical terms this allows us to operate any number of repeater stations on the same frequency. It no longer matters whether we watch the original flag waver or his copycat at the repeater station. If we are unsure of some flag movements we can base our judgment on what both the original and the copycat are doing. Unlike conventional broadcasting the signals do not interfere with but reinforce one another.

European broadcasters have been attracted to this technique because it allows them to extend the geographic coverage of networked programs without the need to find additional frequencies for the repeater stations. Where power outputs are restricted or where the target market has an irregular shape it is more economic to deploy a number of low-power on-channel cellular repeater stations rather than a conventional single high-power transmitter.

## MAGIC NUMBERS

Setting the standards for high definition television must appear a little like a game of chance. There is a range of competing conventional and high definition television numbers to choose from. Which numbers will prove lucky and which ones will lead us astray?

525     The conventional American TV system specifies a total of 525 lines of which 480 lines are visible on the screen. The picture is interlaced and is fully refreshed nearly 30 times a second.

625     The conventional European TV systems specify a total of 625 lines of which 576 lines are visible on the screen. The picture is interlaced and is fully refreshed 25 times a second.

1,050     Two of the four American HDTV systems specify a total of 1,050 lines of which 960 lines are visible on the screen. The picture is interlaced and is fully refreshed nearly 30 times a second. We shall come to the alternative progressive scan HDTV systems in a moment.

1,125  The Japanese HDTV system specifies a total of 1,125 lines of which 1,034 are visible on the screen. The picture is interlaced and is fully refreshed 30 times a second.

1,250  The European Eureka HDTV system specifies a total of 1,250 lines of which 1,152 lines are visible on the screen. The picture is interlaced and is fully refreshed 25 times a second.

We are invited to take our pick.

In practice it is difficult to see much difference between the demonstrations of rival HDTV display systems. When shown on the same high quality display screen the interlaced HDTV pictures from the rival 1,050, 1,125 or 1,250 camps all look much the same—they are all excellent.

What of the pro-scan proponents?

787½  The two pro-scan American HDTV systems specify a nominal total of 787½ lines of which 720 lines are visible on the screen. The picture is progressively scanned nearly 60 times a second.

In the last chapter we discussed a hypothetical contest held between the progressive and interlaced book readers (page 104). It was agreed that the progressive reader derived as much pleasure from a 700-page book as the interlaced, page-skipping, reader might take from a 1,000-page book.

Would the complementary television studies bear this out?

Would viewing tests show that there is little overall perceptual difference between a 1,000 line interlace scan television picture and a 700 line progressively scanned picture?

The results of the American HDTV system tests between the 1,050 line interlaced pictures and 787½ line pro-scan pictures have been inconclusive.

On the one hand the images from the 787½ line pro-scan camera which were sent through the complete digital chain were noticeably worse than the 1,050 line interlaced pictures which were themselves derived from an 1,125 line Japanese HDTV camera. It became clear that like was not being compared with like—the production 1,125 line camera gave "cleaner" pictures than the prototype 787½ line pro-scan camera.

On the other hand the electronically generated test picture of revolving pyramids looked far better through the 787½ line pro-scan chain than through the 1,050 line interlaced chain. Many believe that this unexpected improvement was due to "experimental error" as the interlaced version of the electronic picture—which had been derived from the pro-scan original—had not benefited as much from any "tuning-up" processes as had the pro-scan output.

After much whistle blowing and many cries of "Foul!" from both sides, the result of the contest between 787½ pro-scan and 1,050 line interlaced scan was declared a draw. But this is just what we would expect so let's press on.

Thus there is little to choose in picture quality between the offerings of 1,050, 1,125, 1,250 interlaced and 787½ progressive.

They are all excellent.

This is not the letdown that it might at first appear because we are now offered a wide choice of similar quality display standards. The need to irrevocably choose one particular line number may no longer be so important. As we shall discover later on there may be some other factors that will help us make a choice.

### Improving the normal definition picture

Up to now we have taken the view that a 700 line progressively scanned picture can take the place of a 1,000 line interlaced picture. This is true. We cannot fail to have noticed, however, that many proposals for improved or enhanced television have applied this technique in reverse. They claim, quite correctly, that a progressively scanned picture is 50 percent "sharper" than an interlaced one.

In terms of our book reading analogy, whatever the number of pages in the book, we accept that the progressive reader will enjoy reading it 50 percent more than the page-skipping interlaced reader. Thus a standard 480 visible line NTSC picture can be made to look 50 percent better (the equivalent of a 720 line interlaced picture) just by changing from interlaced to progressive scanning (480 lines + 50% = 720 lines).

The Europeans have demonstrated the same technique.

A conventional 625 line—576 visible line—interlaced picture can be made to look the equivalent of an 864 visible line interlaced picture by switching from interlaced to progressive scanning.

It does not take more than a moment to notice that the 864 lines equivalent interlaced resolution of a 625 line progressively scanned picture is not that much different from the 960 line interlaced resolution of the proposed high definition American 1,050 line standard.

> Many observers, who have been able to see conventional 625 line television pictures redisplayed in the progressive scan format, seriously question the need for anything better.

The 625 line pro-scan pictures look excellent, perhaps not quite "high definition quality" but completely free of the annoying "lininess" of conventional interlaced pictures.

By the same token the Americans consider that progressively scanned 525 line television displays offer an enormous improvement in picture quality over the conventional interlaced format.

There is thus an interesting middle ground between today's conventional definition interlaced 525/625 line television systems and the so called High Definition 1,050/1,125/1,250 line interlaced systems.

Many believe that this middle ground, variously termed Advanced, Improved or Enhanced television, is a more than adequate improvement over what we have today. They believe that it is a good example of the 80–20 rule, that empirical rule of thumb which states that we can often gain 80 percent of the improvement for only 20 percent of the effort.

For the purist there are even higher levels of definition to be obtained. We can play the same trick—switching from interlaced to progressive scanning—on the high definition 1,050/1,125/1,250 line formats. Table 3-2 indicates the improvement which can be obtained. For example, a progressive scan version of the European Eureka 1,250 line interlaced system can offer a definition which is the equivalent of 1,728 interlaced lines.

**Table 3-2**   Interlaced and Progressive Scan Television Line
Systems Expressed in Terms of the Equivalent Interlaced Resolution. The Highlighted Entries Are
Those Chosen by the American Grand Alliance (GA)
for Further Development.

| Nominal No. of Lines | No. of Visible Lines | Type of Scan | Equivalent Interlaced Resolution |
|---|---|---|---|
| 525 | 480 | inter | 480 |
| 625 | 576 | inter | 576 |
| 525 | 480 | prog | 720 |
| 625 | 576 | prog | 864 |
| **787½** | **720** | **prog** | **1,080** |
| 1,050 | 96 | inter | 960 |
| 1,125 | 1,034 | inter | 1,034 |
| **1,125** | **1,080** | **inter** | **1,080** |
| 1,250 | 1,152 | inter | 1,152 |
| 1,050 | 960 | prog | 1,440 |
| **1,125** | **1,034** | **prog** | **1,551** |
| 1,125 | 1,080 | prog | 1,620 |
| 1,250 | 1,152 | prog | 1,728 |

It is relatively easy to become overwhelmed with such a wide
range of choice unless we are able to discover an everyday reference point on which to base our judgment.

It is therefore a welcome surprise to learn that, in the workplace, enhanced and high definition television is already
with us.

## Multi media

Recent improvements in the picture quality of our workplace
computer screens have outstripped the display quality of our
television at home. Not only can we now spend our days seated in
front of a computer screen that is no more than a thinly disguised
enhanced or high definition television set, we can also enjoy the
fruits of the research into the avoidance of flicker.

These days we have become quite used to using a workplace computer screen which has, almost by default, been built to the VGA (Video Graphics Display) standard or better. This standard specifies 480 visible lines on the computer screen. The picture is progressively scanned in the range 60 to 72 times per second.

In recent years the personal computer industry has produced two enhancements to the VGA standard.

- The first super VGA enhancement displays 600 visible lines on the screen.
- The second super VGA enhancement displays 768 visible lines on the screen.

In both cases the picture is progressively scanned and is refreshed in the range 60 through 72 times per second.

Without realizing it we have been introduced to:

Improved NTSC   The 480 line VGA standard provides us with the improvement we would expect if we were able to upgrade the familiar 525 line NTSC television picture from interlaced to progressive scan.

Improved PAL   The 600 line super VGA standard is no more than an example of the similar improvement that progressive scanning would make to the PAL 625 line picture.

You will recall there is a second super VGA standard which is sometimes also referred to as the XGA standard. This standard has enjoyed increasing popularity because of the market success of "Windows," a software program which sets out to emulate the clutter of the desktop on our computer screen. XGA allows us to see the screen clutter in ever finer detail.

XGA is an everyday example of a 768 visible line progressive scan display—remarkably similar to the American 720 visible line pro-scan high definition television proposals and the direct subjective equivalent of Europe's 1,250 line interlaced HDTV system.

How have these improvements come about?

We know that within the space of ten years scientific research first indicated and then demanded that the screen refresh frequency should be raised from 60 to 72 Hz.

We also know that within the space of 10 minutes our personal computer can be opened up and a new display screen "driver card" plugged in to replace the old one.

Hey, Presto! The definition is improved and the flicker has gone.

For a few (hundred) dollars more the display screen itself can also be replaced with a bigger, brighter version of what went before. The best part of this upgrade procedure is that the software programs, which may include our favorite word processor, spreadsheet and drawing packages, need not be changed.

Television engineers have long envied the flexibility enjoyed by the computer screen designers. They asked:

1. How can our television sets enjoy the same flexibility of display that has been built into the personal computer screen?

2. Is it possible to transmit the same, or very nearly the same, conventional TV picture, in order to take advantage of much better picture quality from a more intelligent (and expensive) television set at home?

As we have already learned, the obvious improvement to picture quality is to redisplay the incoming interlaced television picture as a progressive scan and at a higher refresh frequency. Flicker should vanish and the picture should appear 50 percent sharper.

In the past many manufacturers have built prototype television receivers with internal interlace to pro-scan converters but the results have been very disappointing.

This disappointment should come as no surprise for, in the earlier chapter, "Perceptions," we learned that the interlaced (page-skipping) reader had enormous difficulties in re-telling the story if the book was being re-written in front of his eyes. The same difficulty is experienced in the prototype pro-scan television display as the old "plot" on lines 1 and 3 fights for credibility with the revised "plot" on lines 2 and 4.

As we would expect, the static scenes look superb but the moving scenes suffer badly from blurred or double images. This is a pity, because television is all about moving images.

On the other hand it is no longer such a surprise that personal computers are so easy to upgrade—for they are expected to display little more than the relatively static scenes of text and bar graphs.

> Thus it is not practical to upgrade a conventional moving TV picture from its normal interlaced to a higher speed progressive scan form.

Or is it?

Might it be possible to transmit some secondary clues that would inhibit the formation of double images when converting from interlaced to progressive scan? By analogy, the story could be made to make more sense to the interlaced reader if he could be kept informed of the revisions in the story line.

For example, we could alert him to the fact that the scene of the story had been moved from England to America and that the secret agent, who was mentioned on page 3, will make no further appearances.

This would greatly help his understanding of the story and should allow him to be able to retell the story to us later in a more coherent form.

We can apply the same technique to the television picture by sending a list of clues that an intelligent television set can use to improve the quality of the display.

How do we send the clues?

That's easy. In an augmentation channel, of course.

Now we can understand why television engineers have been so willing to take on all the difficulties of digital television—it allows them at last to switch to progressive scanning and to get even with those who design computer displays.


## THE CLASS OF 1993

On April 19, 1993 FCC Commissioner Ervin Duggan addressed the MSTV annual meeting in Las Vegas. He said:

1. To begin with HDTV will be expensive. It will cost a station a million dollars simply to "pass through" a network HDTV signal. The conversion to HDTV should be a transition, not a death march for the broadcaster. Consumers should have the last word on HDTV. Let's leave open an honorable path of retreat if consumers do not embrace HDTV as quickly as we hope.
2. We know that HDTV will be digital. DBS (Direct Broadcast Satellites) will be digital and cable will soon be digital as well. If broadcasters don't enter the digital world I fear you will be left behind in the starting gate. The FCC's own timetable for digital television may be forgiving but the marketplace will not be.
3. We need to encourage the "Grand Alliance." Some fear that we will tilt the standard setting process too much to favor computer technology or other media. We must ensure that our ultimate HDTV standard does not leave broadcasters isolated from the rest of the digital video world. The FCC has endorsed a system that is excellent for broadcasting and that also fits with alternative delivery systems. Perfecting such a system should be one of our goals as we near the HDTV finishing line.

Six weeks later, on May 24, 1993, Richard Wiley was able to announce that a Grand Alliance (GA) had been agreed between the four digital HDTV proponents. The goals of the business and technical agreement were to facilitate interoperability among broadcasting, cable, computer and telecommunications technologies. The Grand Alliance system will use 60 Hz pro-scan and square pixels. All HDTV receivers with screens of 34 inches and above would incorporate a pro-scan display mode. All film material will be transmitted in pro-scan. It is the intention to move to a 1,000 plus pro-scan standard as soon as feasible.

### The quality of 35 mm film

Many of us own a regular 35 mm camera which we dust down once a year for Thanksgiving or some equally important family occasion. We have learned from experience that the best results can only be obtained when the picture we want to record fully fits the frame of the camera's viewfinder. Unlike the old days, when camera film stock was measured in inches not millimeters, any attempt at enlarging a small part of the 35 mm negative ends in tears. The enlarged picture looks fuzzy and the grain of the film, which never caused a problem before, is suddenly only too apparent.

Post production film houses have discovered that 35 mm movie film can suffer the same fate if it is enlarged too far. Their business is a threefold process. The first step is to transfer individual frames of a 35 mm movie from film into the electronic domain. The second step is to manipulate these "ex film" images by means of the many TV post production techniques at their disposal. The third step is to transfer these modified electronic images back onto new 35 mm film stock.

But their business has hit a snag. To what resolution should they scan the film image? As we have discovered "domestic" HDTV standards are aiming at a horizontal definition of 1,920 pixels so surely they must go one better. They have therefore doubled the resolution from 2,000 to 4,000 pixels but the results have been a disappointment. They have discovered that, at this ultra high level of definition, each 10 bit sample of video contains 3 bits of noise.

Where does all this noise come from?

From the grain in the film.

In technical terms the video signal to noise ratio (s/n), which we first met in Chapter One page 24, of the ultra high definition transfer is no better than 40 dB.

When we recall (page 24) that the accepted limit of fringe area performance of a conventional 525/625 line TV picture is set as 43 dB s/n, that our cable TV hookup offers at least 48 dB s/n and that we only perceive the pictures as "solid" when we use a good rooftop antenna that delivers a good quality signal which is above 50 dB s/n then we have a problem.

How can we solve it?

If the post production house halves the definition from ultra high 4,000 pixel to regular high 2,000 pixel then they halve the effect of the grain in the film. The s/n rises from 40 to 46 dB.

Can they do better?

If they halve the definition once again from 2,000 to 1,000 pixels then they halve the effect of the grain once more. Image quality rises from 46 to 52 dB s/n. It is only when we reach this level of quality that we start to perceive the picture as "solid."

> The harsh fact is that 35 mm film is, in numerical terms, not that "high definition."

The idealists argue that 65 mm film will look much better. Failing that, 35 mm film should be permitted to capture bigger images by extending the picture into the space currently assigned to the optical sound track.

The pragmatists argue that it is probably sufficient to implement electronic HDTV systems that offers no more than a thousand pixels of horizontal definition in order to produce "solid" pictures. They believe that the current 787½ line proposal is a good match to 35 mm film and that higher definition systems are a liability until far better film stock is generally available.

To please the film company the post production house will transfer the film at the ultra high 4,000 pixel definition but must take steps to chop out the noise. The simplest way to do this is to reduce the precision of the video sample from 10 to 7 bits.

Hey, Presto, the 3 bits of noise have vanished.

Unfortunately this can easily change the "film look" of the original image into a flatter "electronic look." We first encountered a similar problem in Chapter One on page 15. We recall that chopping the last 4 bits off a 20 bit audio recording in order to fit it on a 16 bit CD produces a gritty sound. In time we can expect that the post production video houses will probably adopt the intelligent dithering technique that is now used in the mastering of audio CDs.

### Square pixels

Up to now we have concentrated on the number of TV lines in a picture, 480 active lines in an American picture and 576 active lines in a European picture. We have assumed that the normal definition 4 by 3 picture will use the 720 horizontal pixels of Recommendation 601. We have also assumed that 16 by 9 normal definition pictures will either increase the horizontal pixel number to 960 or cheat a little and stick at 720 pixels.

There is a body of opinion, however, who would like to see a greater precision in the choice of horizontal pixels. Their argument runs as follows.

An American TV picture with 480 active lines has, in effect a resolution of 480 vertical pixels. If each picture element is to be square then for a 4 by 3 ratio screen format, there should be 4/3 as many horizontal pixels as vertical ones. 4/3 × 480 = 640 horizontal pixels. Cramming more pixels in the line will make each pixel less wide than square. Naughty Rec 601 with its 720 pixels is thus guilty of squashing in too many pixels per line.

However, by the same token, the European picture with 576 active lines has a resolution of 576 vertical pixels. Once again, if each picture element is to be square then the 4 by 3 screen format suggests that there should be 4/3 × 576 = 768 horizontal pixels. Rec 601 now errs in the opposite direction, its 720 pixels are now wider than the idealized square shape.

Few broadcasters have ever lost sleep over these two theoretical disparities. NTSC viewers at home are lucky to see 330 pixels of horizontal resolution—let alone 640, 720 or 768.

But, as we have seen, Rec 601 has gained a following. Normal definition 4 by 3 studio pictures now aim to have 720 pixels, wide-screen pictures are often defined with 960 pixel horizontal resolution and high (double normal) definition pictures have 1,920 pixels.

But what have we done?

In arriving at the figure of 1,920 we have started with the Rec 601's historic international fudge of 720 pixels, expanded it to 960 in order to fill a 16 by 9 wide format screen and then doubled it.

Unfortunately the proponents of the square pixel have used this horizontal value as the starting point for the reverse definition of the vertical resolution. They argue that, for a 16 by 9 format there should be 9/16 as many vertical as there are horizontal pixels. 1,920 × 9/16 = 1,080 vertical pixels.

As a result of these calculations there is now a school of thought that suggests that the 1,080 by 1,920 square pixel picture format will facilitate interoperability between television and computer products. Their arguments are founded on the assumption that computer displays use square pixels. To date this is only partly true.

The super VGA card in my PC offers a number of screen resolutions. There is no doubt that the 640 by 480, 800 by 600 and 1,024 by 768 resolutions are undoubtedly square when viewed on a 4 by 3 shaped screen. There is also no doubt that the other equally valid choices of 640 by 400; 720 by 348, 352, 384 or 512; and 1,056 by 350 or 480 are not square.

Strange as it may seem, neither my eye nor my computer monitor mind these changes at all, a small tweak to the height and width controls allows me to make the display any size I like.

In the last ten years we have seen the introduction of the CGA, MDA, EGA, VGA and SVGA computer display formats. On this track record it is difficult to remain convinced that the personal computer industry has finally settled on the 1,080 by 1,920 format as the answer to interoperability.


## MUTUAL AGREEMENT

It is to the credit of the members of the Grand Alliance that they appear to have been able to satisfy all comers.

The Grand Alliance has adopted three standards.

1.  A pro-scan square pixel standard of 720 active lines by 1,280 horizontal pixels. This so-called 787½ line format will be used to transmit film material.
2.  An interlaced square pixel standard of 1,080 active lines

by either 1,920 or 1,440 horizontal pixels. This will
provide compatibility with the existing 1,125 line interlaced production format.

3. A pro-scan square pixel standard of 1,080 active lines
by 1,920 horizontal pixels. This is intended for future
development and possible multi-media compatibility.

Current thinking is that the first and second options may, in
time, grow into the third.

The figures in the second option are something of a surprise
and beg the following questions.

1. What has happened to the original proposals for a 1,050
interlaced line standard?

2. How does a 1,080 active line standard offer compatibility with the existing 1,125 line production standard?

3. Where does the figure of 1,440 pixels come from?

Let's try to address each question in turn.

**1.** A little earlier (Table 3-2) we discovered that the pro-scan
performance of European 625 line TV was equivalent to an
interlaced picture that contained 864 active lines. The 1,050 line
interlaced proposal uses 960 active lines, an 11 percent improvement over the equivalent performance of a pro-scan 625 line
system. The Grand Alliance was not slow to recognize that pur-
suing a mere 11 percent improvement over Enhanced Definition
European pictures was just not worth the effort. Something better
was needed.

**2.** In Chapter One (page 37) we first learned that old-
fashioned analog TV systems cannot show every line—about 10
percent of the lines never get displayed but are "lost" elsewhere
in the system. The 525 line TV system can only show 480 visible
lines and the now discarded 1,050 line HDTV proposal can show
only 960.

The Japanese 1,125 line HDTV system suffers the same
constraints, only 1,034 or so of the 1,125 lines can be displayed
on the screen.

How can we shoe horn the Grand Alliance 1,080 active lines into the 1,034 lines of the 1125 line HDTV production system?

Do you recall (page 37) that the regular supporters of Rec 601 rarely filled each horizontal line to the brim with 720 pixels. Most times they were content to settle for about 20 or 25 pixels less.

Surely no one would mind if we play the same trick with the horizontal rows of 1,920 high definition pixels by leaving the same percentage of pixels (50 or 60) unused. It is but a small step to push this line of thought a little further by dropping 80 pixels and using only 1,840 of the 1,920 pixels on offer. In this way we can still retain square pixels and achieve compatibility between the GA 1,080 system and the 1,125 production system. The price we have paid for this fudge is the introduction of a thin black band around the edge of the screen. In practice no one will ever know it is there because TV manufacturers always adjust their TV sets to overscan the display screen by a few percent.

**3.** By now we cannot easily forget that a 16 by 9 screen with 1,080 active lines "needs" 1920 horizontal pixels if the pixels are to be square. By the same token a conventional 4 by 3 screen needs 1,440 pixels! Twice the 720 pixels of Rec 601.

Déjà Vu. Recall the earlier European MAC TV family argument whether wide-MAC should be broadcast with the regular 720 horizontal pixels of the Rec 601 standard or whether the wide-screen picture should be enhanced to 960 pixels. We are witnessing another round of conflict between the idealists and the market pragmatists—but this time on American soil.

## Why 787½ lines

We have seen how all analog TV systems are able to show only about 90 percent of the total number of lines on the screen. The remaining lines get "lost" in the high speed "fly back" from the bottom to the top of the picture.

Digital systems do not suffer the same 10 percent loss of display time which is normally expressed in terms of lost lines. In digital television systems it is only the number of active lines that have any true meaning. The published figure for the total number of lines—set about 10 percent higher than the "active" number—is therefore no more than a convenient fiction which has been introduced to ease the comparison with analog TV systems.

In reality the description "787½ lines" is meaningless when applied to a digital TV system. It has come about through a deliberate slip in reasoning.

*Step 1.* The 720 line digital TV system has 50 percent more active lines than the regular 480 active line NTSC television.

TRUE

*Step 2.* If the new digital system has 50 percent more lines than regular 525 total line NTSC TV then logic dictates that the new system must have 50 percent more lines than regular 525 line TV.                                                               FALSE

*Step 3.* The new digital system must therefore have 787½ lines.                                                                 FALSE

The deliberate slip has been introduced in order to save us from any confusion that may arise between the 720 horizontal pixel resolution of Rec 601 and the 720 active lines of the pro scan HDTV system.

### The pro-scan studio

It is generally agreed that the best starting point for generating digitally compressed TV pictures is the use of progressive scanning in the TV cameras and the other equipment, such as telecine, which generate the TV pictures.

### Pro-scan film

Imagine therefore that the broadcaster has read of the advantages of pro-scan and would like to try it for himself.

Can he modify his existing telecine machine?

Yes. The major cost of a telecine machine lies in its film transport. This is designed to hold each frame of the film rock-steady as it is scanned. Modern telecine machines offer considerable flexibility in scanning methods and the modification from interlaced to progressive scanning is relatively easy to accomplish. At the 1993 Las Vegas NAB show one UK manufacturer showed how an older 525 line machine could be modified to provide sparkling 787½ line 60 Hz pictures.

### Pro-scan camera

Can the broadcaster modify his existing television cameras to produce pro-scan rather than interlaced pictures?

The short answer is No.

If we return to the gardening analogy we remember that we had the choice of soaking the garden once every two weeks (pro-scan) or providing a light watering every week (interlaced). If we were able to modify the 525 line camera to provide pro-scan pictures then the effect would be the same as fully watering half the garden one week and fully watering the other half of the garden the following week. The effect is certain—the plants will not survive the longer period without water.

If the camera was modified from interlaced to pro-scan there would still be 30 full pictures a second. They would now be presented, however, in a form that would produce the most appalling flicker.

Thus, within the limits of a 525 line television system and 30 pictures per second the existing cameras cannot be successfully modified to pro-scan.

One way to circumvent this difficulty is to introduce a temporary HDTV service in an interlaced format—the Grand Alliance's second option. However, we shall discover another way around this difficulty in a later section of this chapter.

### A WORLD STANDARD FOR DIGITAL HDTV

When electronic television first superseded its mechanical older brother in the 1930s it was quickly realized that the use of interlaced scanning could greatly alleviate the problem of flicker. The interlace effect could be achieved more easily if an overall "odd" number of lines were chosen as the TV standard. Thus the well-known television line standards of 405, 525, 625, 819 and

1,125 lines were all based on odd number, low value prime factors and thus betray their origins in the need for interlaced scanning.

The frequency of the public electricity supply determined the number of television pictures per second—50 fields (half pictures) in Europe and 60 fields per second in the United States. For 60 years the television picture rates have hardly changed though higher picture refresh rates have been mooted from time to time.

If the higher rates were indeed the improvement they suggest, why were they not adopted?

In order to understand this resistance to change it is useful to divide the television chain into two parts, the generation and the display of the picture. It is easier to deal with them in reverse order.

## Picture display

In everyday life we are likely to watch displays which are generated at different refresh frequencies. In America the domestic television offers 60 fields a second whereas the personal computer screen offers 60 to 80 frames a second. Apart from a difference in our perception of flicker, we notice no ill effects when watching a 70 Hz computer screen in a working environment where the ambient illumination is derived from fluorescent strip lights. We know that this form of lighting is not as steady as it might first appear for it is pulsed at a multiple of the public electricity frequency, in this case 120 times a second.

In Europe, where the fluorescent lights pulse at 100 Hz, no ill effects are observed when working on a computer screen which is itself refreshed at 60 or 72 Hz.

Thus our eyes are relatively immune to any unpleasant cross-effects—more properly termed inter-modulation—which might occur when displays and lighting are driven at different frequencies.

## Picture generation

The TV camera is not as accommodating as is our eye to different refresh frequencies. In America we have all watched "financial" TV programs which include a shot of a dealer's computer screen that seems to flicker very badly. The reason is clear: the American TV camera is set to a standard 60 fields per second whereas the computer screen is set to a higher refresh frequency, typically 70 Hz. The result is quite familiar, the computer screen appears to flicker at 10 "flicks" per second, the arithmetic difference between the two refresh frequencies.

Yet the financial dealer and the television crew appear to see nothing amiss with the "live" computer screen.

More generally we cannot expect to produce satisfactory pictures from a television studio in which the lighting exhibits any residual flicker that differs in frequency from the picture refresh frequency in the camera. If we place a 60 Hz NTSC TV camera in a European TV studio, which uses conventional 50 Hz electricity for the lights, the pictures will be marred by 10 Hz flicker. A 50 Hz PAL TV camera would provide equally unsatisfactory pictures in an American studio with 60 Hz lighting.

There is thus good news and bad news:

- The bad news is that the TV camera refresh frequency must always closely match the electricity supply if a differential flicker effect is to be avoided.
- The good news is that our television display screen is not so constrained. It is not tied to either the frequency of the local electricity supply nor the refresh frequency of the studio camera.

It seems very unlikely that TV cameras will ever move away from a close relationship with the local electricity supply frequency, 50 Hz in Europe and 60 Hz in America. At first sight this would appear to rule out a single picture repetition standard for high definition television. On the other hand, as our improved personal computer displays have already shown, we are able to retain considerable flexibility in the choice of refresh frequencies for the television display at home.

There still remain a number of indirect approaches that might allow us to establish a worldwide HDTV standard that can be seen to be to no one's disadvantage.

We are by now quite familiar with the television field rate refresh frequencies of 50 Hz and 60 Hz. From the earlier discussion of flicker in the second chapter, "Perceptions," we have encountered 72 Hz as the preferred minimum computer screen refresh frequency and have noticed that a 75 Hz display refresh frequency nicely straddles our perception of central and peripheral vision at typical values of TV screen illuminance.

Television research has highlighted both 72 and 80 Hz as candidate values of refresh frequency which are likely to appeal to both the 50 and 60 Hz camps if any future move were to be made towards a single picture refresh standard. Long before the results of the CFF flicker studies became known the value of 72 Hz was proposed as an easy multiple of the 24 Hz movie film frame rate. But, as we shall discover shortly, the value of 72 Hz has some disadvantages for the European broadcaster.

In the USA researcher Kerns Powers was impressed by the BBC idea of an 80 Hz value, lying 20 Hz away from the American 60 Hz standard and 20 Hz away from 100 Hz (double the European 50 Hz standard). Upconversion from 50 or 60 Hz to the proposed new 80 Hz standard offered the politically novel idea of slight but equal discomfort for both camps, for they both will need to eliminate any residual 20 Hz flicker in the picture.

If the goal of a single worldwide picture rate is to be denied us then perhaps the next best option is to choose a number of refresh frequencies that can relate easily to one another.

Which are the best numbers to choose?

Table 3-3 sets out a number of picture refresh frequencies in the range 24 to 120 Hz. Each quoted refresh frequency value is associated with a specific application, viz:

24 Hz  is the movie film picture (frame) rate

25 Hz  is the European television frame rate

30 Hz  is the American television frame rate

50 Hz  is the European television field rate

60 Hz  is the American television field rate

72 Hz  is the computer screen recommended minimum display rate.

75 Hz  is the straddle across our central and peripheral vision

80 Hz  is the "equal discomfort" field rate

**Table 3-3** Relationship between Picture Refresh Frequency, Frame Number and Elapsed Time for Movie Film and Television Systems. The 200 ms Repetitions Are Highlighted.

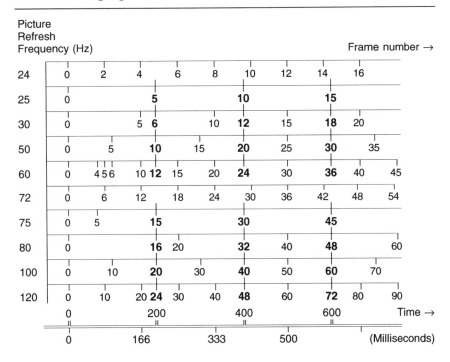

Picture Refresh Frequency (Hz) — Frame number →

| Refresh (Hz) | Frame numbers across time → |
|---|---|
| 24 | 0  2  4  6  8  10  12  14  16 |
| 25 | 0  5  10  15 |
| 30 | 0  5  6  10  12  15  18  20 |
| 50 | 0  5  10  15  20  25  30  35 |
| 60 | 0  4 5 6  10 12  15  20  24  30  36  40  45 |
| 72 | 0  6  12  18  24  30  36  42  48  54 |
| 75 | 0  5  15  30  45 |
| 80 | 0  16  20  32  40  48  60 |
| 100 | 0  10  20  30  40  50  60  70 |
| 120 | 0  10  20 24  30  40  48  60  72  80  90 |
| Time → | 0  200  400  600 |
| (Milliseconds) | 0  166  333  500 |

100 Hz  is the European field-doubled rate
120 Hz  is the American field-doubled rate

Table 3-3 shows that, at regular intervals, the separate frame sequence numbers of the different picture rates will coincide exactly in time. For example, every fifth of a second (every 200 milliseconds) the 25 Hz repetition rate will produce 5 complete frames whereas the 30 Hz repetition rate will produce 6 complete frames.

If we want to convert from a 25 to a 30 Hz picture repetition rate we need to generate six output frames for every five input frames. Although the fifth frame at 25 Hz is identical to the sixth frame at 30 Hz we have no choice but to construct all the intermediate frames of the 30 Hz sequence by the judicious mixing of more than one of the intermediate 25 Hz frames.

This is not easy.

The interpolation process normally requires a great deal of complex digital processing if it is to be achieved without significant picture degradation (blurring). Table 3-3 shows that the 200 millisecond time interval is particularly rich in possibilities for it offers a common cross-connect point for a number of repetition frequencies: viz. 25, 30, 50, 60, 75, 80, 100 and 120 Hz.

If we restrict ourselves to higher picture repetition frequencies we no longer need to interpolate the picture across such a wide time window. A 50 to 60 Hz conversion need take only 100 milliseconds and a 60 to 75 Hz conversion need take only 66 milliseconds. This is because the common factors in the sequence repeat themselves more often at the higher refresh frequencies.

Before the arrival of digital television such a detailed discussion of picture frame rate conversion and interpolation was best left undiscovered, perhaps safely tucked away in the annals of learned broadcasting societies or explained away in the glossy operating manuals that are said to accompany $100,000 television standards converters.

> But the hard digital processing work has already been done for us.

In the first chapter, "Digits Galore," we discussed how the television picture might be digitally encoded into two forms, absolute frames and differential frames. We agreed that we would need to transmit an absolute frame every hundred milliseconds or so but that the intermediate frames might include shorthand instructions such as "move the car a little to the right," "do it once again," etc. Providing that we do not overdo this differential technique the regenerated sequence of moving pictures should appear more than satisfactory.

We may safely assume that the Grand Alliance American designer of a digital HDTV system would expect that his domestic market HDTV receiver will use a 60 Hz picture refresh rate whereas the equivalent European designer will assume a 50 Hz refresh rate for his viewers. However, neither designer expects to receive 60 or 50 absolute pictures a second. It is more likely that only about a fifth of the frames would be in "absolute" form, the remainder of the transmission will contain clues to the inter-mediate, interpolated or differential frames.

As we can determine from Table 3-3, within the space of 200 milliseconds the 60 Hz American HDTV designer will be expected to move the car from A to B in 12 short interpolated steps whereas the 50 Hz European designer must perform the same maneuver in 10 slightly larger steps. Furthermore, if we choose to refresh the television display at 75 Hz, (our central vision/ peripheral vision straddle frequency), moving the car from A to B will require 15 shorter steps.

> Depending on the display refresh frequency, we need to be
> able to interpolate 10, 12 or 15 steps within the 200 mil-
> liseconds time window.

Although this technique allows us to convert easily between European 50 Hz, American 60 Hz and CFF 75 Hz we need the relatively long time window of 200 milliseconds and as many as 10, 12 or 15 pictures "in flight" at any time. Is it possible to reduce the size of the time window and the number of picture frames "in flight" at any time?

Let's look at the problem from both sides of the Atlantic.

**USA** We know that the American television producer will prefer to "shoot" his program material at 60 Hz, for the use of any other lighting frequency is impractical.

The American viewer will expect to watch the program at a display refresh frequency of at least 60 Hz for he has heard that European TV sets flicker very badly. In the USA a 50 Hz refresh frequency is a non-runner.

**Europe** We know that the European television producer will prefer to shoot his program material at 50 Hz for the same practical reason as his American counterpart sticks to 60 Hz. The European viewer will watch at

**Table 3-4**   Changing the Display Frequency.
This Chart Illustrates the Conversion Ratios and the Time Span Needed to Convert from One Display Frequency Standard to Another.

| From \ To | 120 Hz | 100 Hz | 80 Hz | 75 Hz | 72 Hz | 60 Hz | 50 Hz | 30 Hz | 25 Hz |
|---|---|---|---|---|---|---|---|---|---|
| 100 Hz | 6:5 / 50 ms | | | | | | | | |
| 80 Hz | 3:2 / 25 ms | 5:4 / 50 ms | | | | | | | |
| 75 Hz | 8:5 / 66 ms | 4:3 / 40 ms | 16:15 / 200 ms | | | | | | |
| 72 Hz | 5:3 / 42 ms | 25:18 / 250 ms | 10:9 / 125 ms | 25:24 / 333 ms | | | | | |
| 60 Hz | 2:1 / 16 ms | 5:3 / 50 ms | 4:3 / 50 ms | 5:4 / 66 ms | 6:5 / 83 ms | | | | |
| 50 Hz | 12:5 / 200 ms | 2:1 / 20 ms | 8:5 / 100 ms | 3:2 / 40 ms | 36:25 / 500 ms | 6:5 / 100 ms | | | |
| 30 Hz | 4:1 / 33 ms | 10:3 / 100 ms | 8:3 / 100 ms | 5:2 / 66 ms | 12:5 / 166 ms | 2:1 / 33 ms | 5:3 / 100 ms | | |
| 25 Hz | 24:5 / 200 ms | 4:1 / 40 ms | 16:5 / 200 ms | 3:1 / 40 ms | 72:25 / 1 sec | 12:5 / 200 ms | 2:1 / 40 ms | 6:5 / 200 ms | |
| 24 Hz | 5:1 / 42 ms | 25:6 / 250 ms | 10:3 / 125 ms | 25:8 / 333 ms | 3:1 / 42 ms | 5:2 / 84 ms | 25:12 / 500 ms | 5:4 / 166 ms | 25:24 / 1.0 s |

Hz or at greater than 72 Hz—where the flicker stops.
Thus 60 Hz may be a non-runner in Europe.

We may assume that the production standards are fixed: 50
Hz in Europe and 60 Hz in America. On the other hand the display
refresh standards could be 50, 72, 75 and 80 Hz in Europe and 60,
72, 75 or 80 Hz in America. Table 3-4 sets out the conversion ratios
and the time window which is required for the same wide range
of picture refresh frequencies that were set out in Table 3-3. Of
particular interest are the conversions from 50 or 60 Hz to 72, 75
and 80 Hz.

**USA**     The conversion from 60 Hz to 72 Hz is a little more difficult than 60 to 75 Hz which, in turn, is a little more difficult
than 60 to 80 Hz.

**Europe**  The conversion from 50 Hz to 72 Hz is particularly
cumbersome whereas the conversion from 50 Hz to
75 Hz is very straightforward.

We may discard the 72 Hz refresh frequency on the grounds
that the 50 Hz to 72 Hz conversion, a ratio of 25 to 36, is too
cumbersome for practical use. The choice between 75 or 80 Hz is
less clear cut: 50 Hz to 75 Hz is very much easier than 50 Hz to
80 Hz whereas 60 Hz to 75 Hz is only a little more difficult than
60 Hz to 80 Hz. Perhaps we may obtain further guidance in our
choice of refresh frequency by a closer reference to the number of
television lines in the European and American HDTV proposals.

## CONVERTING THE LINE STANDARDS

In the introductory section of this chapter we established
that all the various international HDTV proponent systems offered us excellent pictures and the only possible difficulty with
this judgment was that we appeared to be spoiled for choice.

However, if an HDTV format is to be commercially successful, it must prove easy to convert between the HDTV system and
today's 525 and 625 line television systems. The European 1,250
line (twice 625) and American 1,050 line (twice 525) HDTV

proposals were an early recognition of this need for an easy conversion.

If the HDTV programs are to reach a worldwide audience then it is important that American productions can play on European TV sets and vice versa. If more than one HDTV system is to survive in the marketplace then it is also important that conversions between rival HDTV systems are easy to accomplish. In short, the different conventional and HDTV formats must bear an easy arithmetic relation to one another, both in line number and picture display frequency.

Table 3-5 sets out the (visible) line ratios between conventional 525/625 line television and the HDTV proponents of GA 720/787½, US 960/1,050, NHK 1,034/1,125 GA 1,080 and European 1,152/1,250 lines. The reason for the American popularity of 787½ lines is soon apparent for it is an arithmetically easy 3 : 2 up-conversion from the NTSC 525 line standard.

In its proposed pro-scan format the 720/787½ line system is the equivalent of an 1,080 active line interlaced system with a fictional total of 1,180 lines. Yes, the Grand Alliance 720 active line pro scan and 1,080 active line interlaced systems are the two sides of the same coin. Moreover, with impeccable political correctness, the Grand Alliance has arranged that their interlaced version falls neatly between the HDTV offerings from Japan (1,125 lines) and Europe (1,250 lines). Table 3-5 shows that, with conversion ratios of 19 : 18 and 27 : 26 no more than a whisker

**Table 3-5**  The Inter-Relationship between the Visible Line Ratios of the Major Television Systems.

| European 1,152/1250 | | | | | | |
|---|---|---|---|---|---|---|
| 19 : 18 | GA 1,080/1,180 | | | | | |
| 12 : 11 | 27 : 26 | NHK 1,034/1,125 | | | | |
| 6 : 5 | 9 : 80 | 11 : 10 | US 960/1,050 | | | |
| 8 : 5 | 3 : 2 | 13 : 9 | 4 : 3 | GA 720/787½ | | |
| 2 : 1 | 15 : 8 | 11 : 6 | 5 : 3 | 5 : 4 | PAL 625 | |
| 12 : 5 | 9 : 4 | 11 : 5 | 2 : 1 | 3 : 2 | 6 : 5 | NTSC 525 |

now separates the three standards. At this point it is surely easier to agree on a single standard than to attempt to convert between the three.

Although the 787½ line system is an attractive up-conversion from NTSC 525 lines the choice of 787½ lines for the Europeans is not so obvious. Table 3-5 shows that the conversion from HDTV 787½ lines to conventional 625 lines requires a 5 : 4 ratio.

But wait a moment, these ratios of 5 : 4 and 3 : 2 have a familiar ring to them.

If we correlate the conversion details in Table 3-5 for 525, 625 and 787½ line television systems against the details set out in Table 3-4 for the conversion ratios of the 50, 60 and 75 Hz picture refresh rates we note that:

| America | Lines | 787½ ↔ 525 | is the ratio 3 : 2 |
| | Refresh | 75 ↔ 60 | is the ratio 5 : 4 |
| Europe | Lines | 787½ ↔ 625 | is the ratio 5 : 4 |
| | Refresh | 75 ↔ 50 | is the ratio 3 : 2 |

It is now quite plain to see that the American and European conversion ratios are the same, the only difference is that the Americans and the Europeans need to apply them differently. The 3 : 2 ratio is relatively easy to achieve whereas the 5 : 4 ratio requires more complexity in the digital circuits. It is, however, a neat compromise, not least because it meets the Kerns Powers political criteria of equal comfort and discomfort to both camps.

## A WORLDWIDE HIGH DEFINITION TELEVISION DISPLAY STANDARD

We have explored the rival high definition standards in some depth. In an ideal world it might prove possible to establish a single worldwide HDTV standard for both production and display but that goal is beset by the presence of two electricity supply frequencies throughout the world.

We are left, however, with a most valuable prize, a flicker-free HDTV display standard which is compatible with conventional and high definition American and European television systems.

The prize is irresistible: A universal 787½ line 75 Hz progressive scan high definition digital television distribution standard.

## Implementation

The current American all digital HDTV 787½ line 60 Hz proposals would need to be changed only slightly in order to allow a 75 Hz interpolation as well as the 60 Hz interpolation of the picture. We know that, in order to generate a 75 Hz picture, we need to interpolate 5 frames for every 4 frames that occur at 60 Hz. By the addition of a few extra clues in the digital television transmission we can thus provide good pictures at both 60 Hz and 75 Hz.

We have seen that the European 1,250 line 50 Hz interlaced HDTV system offers virtually the same picture quality as the American 787½ line pro-scan display. Conversion from 625 lines to 787½ lines requires a nontrivial 5 : 4 conversion whereas the up-conversion from 50 to 75 Hz is more straightforward.

For a long time to come the major source of HDTV programming promises to be film. Are there any further tricks we can play in order to effect a good match between the two media?

### 35 mm movie film frame rates on HDTV

Hollywood usually shoots its movie film at 24 frames per second. Thirty years ago the television studio used a telecine machine, which is a combination of a movie projector and television camera, to show us the movie on our television set at home. Nowadays the TV studios are more cautious. They know that the fragile celluloid film can easily break at the wrong moment so they first transcribe the film onto videotape. It is the videotape playback that we see at home.

In Europe it is standard practice to run the telecine at 25 not 24 frames a second in order to dovetail into the 25 frames per second television standard. No one minds the film running

4 percent faster, no one, that is, except those with perfect pitch who are perplexed to discover that the film music has risen unexpectedly two thirds of a semitone in pitch. Even this change in musical pitch can be corrected by the use of the "musical" word processing techniques which were described in the first chapter.

Although the American telecine is set to play back the film at its original speed, 24 frames a second, it is unable to effect an entirely satisfactory match with the 30 pictures per second NTSC television standard. Some fields of the television picture offer good sharp definition whereas other fields contain a dumb mix, a "smudge," of two adjacent frames of the movie film. In technical terms this smudge occurs because of a strange effect called "The Three-Two Pull-down" which is described in schematic form in Table 3-6.

We recall that regular 525 line interlaced TV pictures are made up of a sequence of odd and even fields which repeat every 1/30 of a second. Each field, whether odd or even, lasts 1/60 of a second. If we look at line A and line B of Table 3-6 in detail we can see that the first frame from the movie film, marked A1 in line A,

**Table 3-6**    The Three-Two Pull-Down timing Protocol.

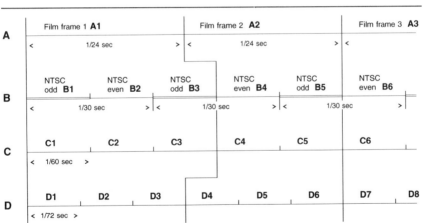

Key: Line A  Movie film—24 film frames per second
     Line B  NTSC TV—60 TV fields per second
     Line C  GA 787½—60 Pro-scan TV frames per second
     Line D  GA 787½—72 Pro-scan TV frames per second

lines up with an odd and even field, B1 and B2, of the NTSC TV system which is shown in line B. Together the odd and even television fields make up a complete TV picture—confusingly called a TV frame. Fine so far.

Now comes the snag.

The second frame of the movie film, marked A2, is not ready to be shown until the next odd field, B3, is already half way down the screen.

What shall we do?

Within the telecine machine it is mechanically impractical to change from film frame A1 to frame A2 while the B3 TV raster is still in flight down the screen. Instead we have to wait until the B3 odd field has finished. It is only at that point in time that we can mechanically "pull down" the film in the telecine machine in order to update film frame A1 with film frame A2.

The next transition, from film frame A2 to film frame A3, is better behaved but we hit the same pull-down problem at the subsequent transition between film frames A3 and A4.

Thus film frames A1, A3, A5 etc. always line up with 3 fields of the regular 525 line TV signal and film frames A2, A4, A6 etc. line up with 2 fields. Hence the strange title "The Three-Two Pull-Down."

Problem solved?

Not at all.

The TV frame that is made up of the odd field B1 and the even field B2 is fine. It is the second TV frame, made up of TV fields B3 and B4 that causes trouble.

The table shows that the odd TV field B3 is taken from film frame A1 whereas the even TV field B4 is taken from film frame A2.

This is just what our interlaced reader in Chapter Two has always dreaded—the old plot on the odd pages of his book being mixed with the new plot on the even pages of his book.

The result of this mix-up is only too well known. The quality of the NTSC telecine transfer from movie film is considerably worse than for PAL transfers where the three-two pull-down artifact does not exist.

One neat way around this difficulty has been for TV production companies to shoot the original film at 30 not 24 frames per second. This works fine for a production which is only intended for American TV distribution but precludes its use in movie theaters or on different TV systems elsewhere in the world.

The Grand Alliance 787½ proposal offers us a better long-term solution.

The GA 787½ system uses progressive scanning at a 60 Hz refresh rate. Its operation is shown in line C of Table 3-6. Although the telecine still employs the infamous three-two pull-down protocol it is now used to align TV frames not TV fields. Thus TV frames C1, C2 and C3 line up with film frame A1 and TV frames C4, C5 line up with film frame A2. There is no unwanted mixing of old and new plots so the quality of the moving picture is very much improved.

However, there is still another improvement to be made.

The three-two pull-down protocol has introduced an unwanted syncopation into the TV's motion portrayal of the movie film. When the movie was first shot the film ran through the camera at a steady 24 frames per second. When it is displayed on television, however, the three-two pull-down protocol spits out an uneven sequence of movie frames that alternate in duration from a 1/20 to a 1/30 of a second.

How can we avoid this?

The easy way around this difficulty is shown in Line D of Table 3-6.

If the local refresh frequency of the television display is raised from 60 to 72 Hz then the unwanted syncopation effect is removed. TV Frames D1, D2 and D3 line up with film frame A1 and frames D4, D5 and D6 line up with film frame A2.

Is this easy to do?

In the United States movie films are always shot and displayed at 24 frames per second. This is very good news for any digital picture compression system—for there cannot be more than 24 changes in the picture every second. The digital TV system will be only too delighted to deliver just that—24 new pictures a second.

It is then entirely up to the local intelligence in the display device, our new high definition TV set, to turn these pictures into a TV display of either 60, 72 or 75 TV frames a second. The same amount of electronics are needed to do all three speeds.

For movies there is thus no effective difference between a Grand Alliance 787½ line 60 Hz system and a Grand Alliance 787½ line 72 Hz system.

Can we push this up the last inch to 75 Hz?

For the European broadcaster the proposed 787½ line 75 Hz HDTV display standard will provide an easy match to his telecine which is always run at 25 frames per second. Just like the American 24 to 72 Hz interpolation, an up-conversion from 25 to 75 Hz poses no movement-related difficulties because the picture movement is frozen on each frame of the film in increments of 1/25 of a second.

Can we tempt the American broadcaster to raise his telecine speed to 25 frames per second in order to comply with a worldwide flicker-free 787½ line 75 Hz HDTV standard? Before the American broadcaster would agree to such an upgrade he must be assured that his other audience, the massive NTSC 525 line 60 Hz audience will be properly catered for.

So, let's recap.

We have just learned that, for movies, it is just as easy to run the domestic display at 72 Hz as at 60 Hz. This increase in speed

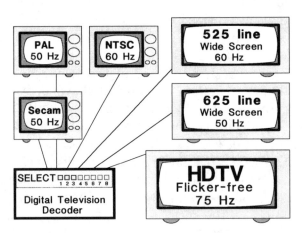

**Figure 3-2**  From the same transmission the omni-standard digital decoder can provide regular and wide-screen NTSC and PAL pictures alongside flicker-free high definition pictures.

avoids the unwanted syncopated motion of the three-two pull-down protocol.

A few pages back we also discovered that, for more general programming, we could add in a few extra "differential" clues to the digital HDTV signal in order to allow 75 Hz interpolation as well as 60 Hz interpolation.

Why not look at this the other way around?

Why not make the 75 Hz interpolation the primary differential signal and add in some extra 60 Hz interpolation signals where required. This change of emphasis allows us to directly address two quite separate markets:

1. The 75 Hz interpolation signal is primarily intended to give us the best possible quality pictures from our brand new expensive High Definition TV sets.
2. The 60 Hz interpolation signal is primarily intended to drive our conventional NTSC 525 line television set which has been relegated to the kitchen or to drive the NTSC video cassette recorder in the den.

It is our second set, the 10-year-old PAL or NTSC television receiver whose picture remains superb but whose cabinet looks a little tired, that holds the key to the success of digital television. We cannot expect these old friends to be turned out of the house and abandoned without further thought for they have brought the world to our doorstep. One of the aims of digital television broadcasting must be to breathe new life into these old friends and to provide us with an even better television service.

Once we have leveled the playing field for the competing media the rest of the chapter offers an explanation of how this might be done.


COMPETING MEDIA

In the introduction to this chapter we learned that some discord was expressed between the over-the-air terrestrial broadcasters and their rivals: cable television, direct broadcasting satellites and videocassette rental. Since then there has been close cooperation between the terrestrial and cable broadcast interests, not least because any off-air signals must be able to pass easily through cable TV networks.

Although the FCC 1988 decision was to concentrate research effort into advanced television for terrestrial broadcasting, the other media have not gone away. Many non-terrestrial service providers took the view that they would try to establish "better" standards if the final ATV terrestrial decisions proved too much of a compromise between picture quality and bandwidth.

In setting the standards for advanced terrestrial broadcasting it would therefore seem worthwhile to expend some effort in making sure that other media are not gratuitously inconvenienced. This section sets out to discover how we might accommodate the views and needs of the other distribution media.

### The telephone company

In the brave new world of multichannel digital high definition television we still need the telephone company to route the television signals around the country and around the world. In the first chapter we learned that there is a hierarchy of telecommunications transmission systems that link our cities and countries together. It is therefore important that, in our choice of digital television transmission speed, we choose values that can easily dovetail into the appropriate hierarchical telecommunications transmission level.

The American digital ATV proponents have chosen the ball park. Their initial research suggests a HDTV video data speed of some 17 to 19 Mb/s to which must be added stereo sound and other data signals. When additional error correcting signals and other housekeeping information are added to these video and audio signals the gross data rate can exceed 25 Mb/s.

We recall from the first chapter that the European telephone companies have established a transmission hierarchy of 2, 8, 34 and 140 Mb/s whereas the American hierarchy is the sequence 1.5, 6, 45 and 140 Mb/s. We can immediately put aside the slower, lower levels in both hierarchies and concentrate our attention on the 34 and 140 Mb/s speeds in Europe and the 45 and 140 Mb/s speeds in America.

> **European 34 Mb/s.** In round figures: 32 Mb/s is assigned to telephone traffic, the remaining 2 Mb/s is used for housekeeping.
>
> **American 45 Mb/s.** In round figures: 43 Mb/s is assigned to telephone traffic, the remaining 2 Mb/s is also swallowed up in housekeeping.
>
> **American and European 140 Mb/s.** The 140 Mb/s link normally contains four 34 Mb/s tributaries plus some extra housekeeping though three 45 Mb/s tributaries could be accommodated in its place.

We are faced with some very interesting choices:

1. **34 Mb/s.** If we limit the gross digital HDTV speed to 16 Mb/s (rather than the Table 3-1 proposals of 21 to 27 Mb/s) then we can fit two HDTV programs into one 34 Mb/s circuit. If the digital HDTV speed should rise above 16 Mb/s then we can only fit one program into the circuit and the cost of transmission over this circuit will double.

2. **45 Mb/s.** If we limit the gross digital HDTV speed to 14 Mb/s then we can fit three HDTV programs into one 45 Mb/s circuit. If the gross digital speed lies in the range 15 to 21 Mb/s then we can fit only two HDTV programs into the 45 Mb/s circuit. Above 21 Mb/s only one HDTV program can be fitted in.

3. **140 Mb/s.** There are four choices:

   14 Mb/s.  If we limit the gross digital HDTV speed to 14 Mb/s then we can squeeze nine HDTV programs into 140 Mb/s (three programs in each of three 45 Mb/s tributaries).

   16 Mb/s.  If we limit the gross digital HDTV speed to 16 Mb/s then we can squeeze eight HDTV programs into 140 Mb/s (two programs in each of four 34 Mb/s tributaries). (See Figure 3-3.)

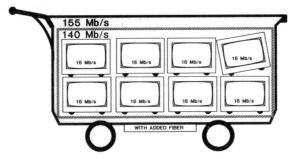

**Figure 3-3** Filling the 140/155 Mb/s supermarket trolley—the telephone company's virtual container can easily accommodate eight 16 Mb/s HDTV programs.

21 Mb/s.   If we allow the gross digital HDTV speed to rise to 21 Mb/s then we can only fit six HDTV programs into the 140 Mb/s circuit (two programs in each of three 45 Mb/s tributaries).

27 Mb/s.   If we allow the gross digital HDTV speed to rise from 21 to 27 Mb/s then we can only fit four HDTV programs into the 140 Mb/s circuit (one program in each of four 34 Mb/s tributaries).

At first sight it would appear that we might squeeze more than these numbers of HDTV programs into the 140 Mb/s circuit—for there certainly appear to be a number of housekeeping areas which might be put to better use. But the telephone companies have been caught in this trap before. They have learned by hard experience that it does not pay to over do it or, by analogy, to pack too much into the 14-foot trailer.

If we are to envisage a scenario in which the telephone company is permitted to provide television service direct into the home then it is probable that, if a digital link is employed, the data speed would be the universal 140 Mb/s standard rather than a choice of 34 Mb/s in Europe or 45 Mb/s in America. In today's technology there is little difference in cost between the provision of 34, 45 or 140 Mb/s circuits for they all need to use exactly the same fiber optic cable to join the two ends of the circuit together.

If the implementation is to be modeled on the more recent Synchronous Digital Hierarchy proposals, the SDH, then the transmission speed rises from 140 to 155 Mb/s though the effective traffic throughput remains the same.

However, there is now a distinct possibility that, in some implementations, the overall data rate may fall.

### Asynchronous transfer mode—ATM

Very few telephone customers are likely to make full use of a 140 Mb/s bulk service all the time. The telcos believe that they can now offer their customers a better arrangement in which the customer can rent—at retail not bulk prices of course—bandwidth on demand. The telcos have readied their new product—ATM.

ATM can offer a net data throughput of any data speed from zero to a maximum of about 110–120 Mb/s, the rest of the 140/155 Mb/s pipe is filled with the telcos' own housekeeping information.

At first sight ATM looks an ideal way to provide a "Video dialtone" service. During business hours the ATM service can carry lots of low datarate videoconferencing calls, in the evening it can target the residential market and carry a smaller number of high datarate HDTV programs.

Difficulties can arise, however, when there is either a lot or a mix of traffic. As we shall discover later on the temptation of statistical multiplexing—our Chinese banquet protocol—may mean that video dialtone services will move away from fixed data rates. If this does occur the video service may occasionally need higher data rates or may be called on to exchange housekeeping information with the telco's ATM switch. This will reduce the net data carrying capacity of the circuit momentarily and may result in either a slight time delay to or a complete loss of part of the data. Let's be optimistic and assume no data is lost.

To a multi-media computer user this hiccup in the ATM service is probably of little significance—for a fraction of a second's delay in the output of his printer is unlikely to be noticed. However, such a delay in a video service might be catastrophic. As we learned in Chapter One the video signal has no time to spare.

Much effort is currently being devoted to the design of "robust" video coding schemes which will not fall prey to time warps or data errors if and when the capacity of the inter-telco or customers' ATM line capacity is unable to meet peak demand.

In order to mitigate these shortcomings the Grand Alliance has suggested that parts of the video data stream should be marked up in advance as suitable candidates for sacrifice when routed through an ATM switch.

Let's get back to bulk distribution.

How many programs are we likely to transmit down the telephone wire?

As we have just learned, the 140/155 Mb/s circuit can support four, six, eight or nine HDTV programs. For a telephone company whose everyday business is the routing of multiple circuits, the natural number to choose from these three alternatives are the "binary" values four and eight.

Let's run the numbers again.

Eight programs through a 140 Mb/s circuit implies a maximum data rate per HDTV program of 16 Mb/s. Four programs would be allowed up to 32 Mb/s each.

It will be tough but can we try for eight, each coded at 16 Mb/s? Each 16 Mb/s digital HDTV signal must contain both sound and vision. Let us tentatively divide this as follows: 14 Mb/s for the video and the remaining 2 Mb/s for stereo sound, teletext and other services.

Can it be done?

Yes and no.

The four 1991 proponents (Table 3–1) suggested two sets of net video data rates.

Pro-scan    The 787½ line pro-scan proposals required 16.92 and 18.88 Mb/s in order to transmit a picture of 720 by 1,280 pixels.

Interlaced  The 1,050 line interlaced proposals required 17.47 and 17.70 Mb/s in order to transmit a picture of 960 by 1,480 and 1,500 pixels respectively.

At the NAB Las Vegas show in 1993 (page 66) a UK company demonstrated a "difficult" Rec 601 quality horse racing video that was coded at 8 Mb/s. The interlaced pictures contained 576 by 700 pixels. Less demanding picture material could be coded at the lower rate of 6 Mb/s. A year later, at the 1994 Las Vegas show, there was little doubt that "difficult" Rec 601 pictures could be coded at 6 Mb/s. and that most material could be coded at less than 4 Mb/s.

> Pixel for pixel it does appear that the Grand Alliance 787½ line pro-scan pictures can be coded into a 14 Mb/s data stream.

However, it may prove too difficult to code a full 1,080 by 1,920 pixel picture—whether interlaced or pro-scan—into that data rate and a higher rate will be needed. Hence the attraction and inclusion of the intermediate 1,080 by 1,440 pixel format as the Grand Alliance interlaced alternative to the 787½ line pro-scan proposal.

Perhaps the most welcome surprise to emerge from all this research is that pro-scan pictures can be coded at the same rate as interlaced pictures. But surely the fully watered pro-scan garden requires twice as much data as the barely watered interlaced garden? In practice it has been found that coded interlaced pictures are appallingly blurred unless we send additional clues about the changes in plot. By the time the secondary clues are added to the data stream it is just as easy to send pro-scan pictures in their place.

Thus, if we are to watch 787½ line HDTV programs that will, inevitably, be routed via telephone company fiber cable, we will need to specify a data rate of no more than 16 Mb/s per HDTV program.

## The terrestrial broadcaster

The terrestrial broadcaster would not disagree with these calculations of net video data rates. He is able to rent either a part or the whole of a 140 Mb/s circuit from the local telephone company to connect his studio to his transmitting antenna some miles away. Although he may assume with confidence that the fiber cable is virtually error-free, he is decidedly less confident about the vagaries of the over-the-air link from his transmitter to

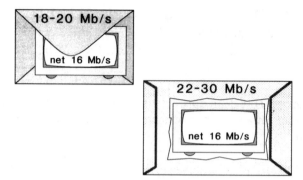

**Figure 3-4** Wrapping up the signal against harm. The terrestrial
wrapping adds some 2 Mb/s of protection against errors
whereas the satellite wrapping adds even more. The
gross terrestrial data rate is about 18 Mb/s, the satellite data rate is about 24–30 Mb/s.

our homes. He needs some further assurance that his signals will
reach us intact.

The 1991 digital proponents covered this in their submis-
sions to the ATV testing center. They proposed that the over-the-
air data rate should be raised by a certain percentage in order to
include some error checking and error correcting signals that both
envelop and protect the HDTV sound and vision data.

At a net rate of 16 Mb/s for the HDTV video and sound it
seems probable that the "enveloping" will add a further 2 to 4 Mb/s
of overhead to the terrestrially broadcast data stream, thus setting the gross terrestrial rate at about 18–20 Mb/s. This is shown
in Figure 3-4. The 1991 US proponents offered overall speeds of
21 through 26 Mb/s and the Grand Alliance has proposed an
overall speed of 27 Mb/s.

### The satellite broadcaster

The satellite broadcaster will also welcome the 16 Mb/s net
HDTV digital transmission standard because he, too, can rent a
part of a 140 Mb/s circuit to carry as many as eight programs from
his studio to the satellite "earth station" that beams his signals
up to the satellite.

He is also concerned about any transmission errors that may
occur in the link from his earth station through the satellite and

into the dish in our back yard. His technical research may indicate that a quite different enveloping procedure is needed to protect his "satellite" signals effectively from harm.

For historic reasons satellite transponders (repeaters) have a channel bandwidth of either 72 MHz or subsets of that amount. Until a few years ago most regular satellite TV channels used transponders with a bandwidth of 36 MHz. Newer satellites now use transponder bandwidths of 24 MHz in order to cram in more channels.

For some time telephone companies have used these 36 MHz satellite transponders to provide international circuits at the standard 45 Mb/s data rate. The same technology allows the 24 MHz transponder to support a 30 Mb/s data stream.

In order to protect the signal against error the satellite broadcaster will elect to provide a very thickly padded envelope around the HDTV data, perhaps raising the gross data rate from 16 to as much as 24 or 30 Mb/s. This is shown in Figure 3-4.

He must first chose whether to

1. Squeeze two HDTV transmissions into each 36 MHz transponder, thus limiting the data rate of each service to just under a half of 45 Mb/s, or
2. Transmit a single 30 Mb/s HDTV service on each 24 MHz transponder.

His decision will be influenced not only by the market price of 36 and 24 MHz transponders but also by his obligations to the cable operator.

## The cable television operator

The cable TV operator must feel unwanted and unloved. The terrestrial broadcaster delivers over-the-air signals direct to millions of homes and the satellite broadcaster can now make use of high-power satellites to do the same. On the other side of the fence he is threatened by the telephone company who would like to take away his livelihood by providing its own rival cable television service.

Yet these threats are not quite what they seem.

We have seen how the telephone company can carry typically eight HDTV signals through a standard 140 Mb/s cable. If these signals can be routed around the world with no loss of quality then surely they can be carried the last few hundred yards from the local telephone exchange to our doorstep?

Yes, they can. The good news is that the telephone company fiber cable can provide us with eight HDTV programs. The bad news is that it is limited to eight.

> In effect the telephone companies' digital technology is too good for the job at hand.

Agreed it can carry a signal around the world without any mishap but the cable television business requirement is much less demanding. The cable TV wire needs to convey the signal no more than a few hundred yards from the roadside distribution cabinet to our home. We know that today's cable TV wire is inexpensive and can carry many tens of American 6 MHz or European 8 MHz TV channels the last few hundred yards without harm.

> At a philosophical level the cable TV and telephone businesses are quite different.

Cable has developed as a broadband medium where we can each pick a different favorite TV program from an offering of perhaps 50 or 100 competing channels. To the telephone company engineer this approach must appear a gross over-provision. He would argue that we are only able to watch one or two channels at a time so why not design a system that can provide a choice of any one channel out of a possible range of thousands. In other words he offers us the television equivalent of a telephone exchange.

In the UK ten years ago this "TV telephone exchange" philosophy initiated a heavy and heady investment in new cable television systems. The marketplace soon discovered, however, that this "switched star" technology was an overkill. Most householders chose to subscribe to only the regular or basic tier package of channels.

It soon proved pointless to tie up further capital in switches that did no more than alternate between the popular "must-carry" free channels—BBC1 and ITV. A hybrid approach, in which all the popular channels bypassed the switch, might be far less expensive.

So the switched star systems were redesigned.

The switched star cable wire, which formerly carried only one or two TV channels out of a much wider range, now carries ten or more. The first four of these channels were quickly spoken for, these were the "must-carry" four national terrestrial channels. The rest of the broadband spare capacity was soon taken up by popular satellite TV channels that were provided "free" in the regular cable TV package. With this redesign of the system the regular cable TV subscriber had no reason to make any further use of the much touted "switched service" so his individual subscriber switch module in the roadside equipment cabinet was taken away. He never knew it had gone.

Experience has shown that the basic mass-market or "broadcast" television program services are best provided by a broadband distribution system—whether via cable or off-air. However, broadband cable does not preclude the addition of an overlay of switched star services for those who can be tempted to pay a premium for specialist or "narrow-cast" television programming.

The cable TV operator takes his programming from local and distant terrestrial transmissions, satellite broadcasts and videotape playout. In America the cable systems are configured to cover a VHF and UHF frequency spectrum which is divided into 6 MHz increments. We recall that the ATV proponents were requested to design their terrestrial HDTV systems to operate within the standard 6 MHz bandwidth allocation so that the proposed digital HDTV signals can pass through a cable TV system as easily as a standard NTSC transmission without the need for any re-engineering.

The satellite television broadcaster is not so fortunate. Satellite television uses a transmission technique—FM—that trades off bandwidth for signal strength. Unlike terrestrial broadcasting, there is rarely any signal strength to spare—at 22,000 miles away from the satellite everyone's satellite TV dish is in the fringe area. As we have just learned the satellite television transmissions normally use up about 24 or 36 MHz of bandwidth, an amount of frequency spectrum resource that is of little consequence provided that our television set is not too far from the satellite receiving dish in the back yard. If, however, the satellite signal is to pass through a cable TV system then the cable operator is justifiably concerned as to the large amount of bandwidth (24 or 36 MHz) that just one TV channel might take up.

Until recently this requirement for a wide bandwidth posed no difficulty as the satellite TV signals were no more than a re-presentation (a re-modulation) of American NTSC or European PAL off-air terrestrial signals. A standard satellite television receiver at the start or the "head-end" of the cable TV system could easily convert the incoming wide bandwidth (24 or 36 MHz) satellite signal into a conventional 6 MHz NTSC or 8 MHz PAL TV signal which could then be sent on its way through the cable system.

European HD-MAC broke the rules.

The satellite part of the MAC TV signal's journey from the studio to our home is little different from the PAL or NTSC signals; HD-MAC consumes the same 36 MHz of satellite bandwidth. Difficulties only arise at the cable TV head-end because the MAC signal cannot be reduced to fit the conventional 6 or 8 MHz bandwidth allocation of NTSC or PAL TV. As we have seen earlier, the cable TV system needs to provide a bandwidth of some 12 to 18 MHz in order to accommodate each MAC channel.

We must avoid making the same mistake again.

When digital high definition television programs are transmitted through high-power television satellites the digital signal will require a similar 24 or 36 MHz bandwidth as its predecessors. As we have seen, the net 16 Mb/s digital HDTV signal will require a protective envelope that may increase the gross transmission rate to a value of anywhere from 22 to 30 Mb/s.

It is then the task of the satellite transmission engineer to design a modulation system, a semaphore code, that permits the transfer of either two 22 Mb/s data signals in a satellite transmitter bandwidth of 36 MHz or a 30 Mb/s data signal in a 24 MHz bandwidth. This is not difficult to do for the textbooks and manufacturers' catalogs are full of suitable techniques that can achieve this level of performance.

We must not forget, however, that the terrestrial broadcast engineers have been just as diligent in designing a quite different modulation system that is able to pack their digital HDTV signal into a terrestrial channel. Their problem is both different and more severe; they must pack a net 16 Mb/s, gross 18–20 Mb/s data signal into a 6 MHz bandwidth. This is new territory and the textbooks can provide only limited help.

By analogy the satellite designer might choose a two-flag semaphore system whereas the terrestrial designer might choose an eight-flag system. From the earlier semaphore example in the first chapter we remember that the two-flag system used the 12 o'clock and 6 o'clock positions whereas the eight-flag system started at 12 and moved on to 1:30 and 3 o'clock, etc. Our two-flag example might thus be a subset of the eight-flag system as both systems use the 12 and 6 o'clock positions.

But it might not prove so easy.

We must accept that, unless the satellite and terrestrial design team efforts are coordinated, the satellite and terrestrial modulation schemes may be incompatible. For example, the terrestrial designer might decide to start his eight-flag signals at the 12 o'clock position but the satellite designer might find good reason to set his two-flag sequence at the 1 o'clock and 7 o'clock positions. The consequence is clear: the satellite and terrestrial systems are no longer related.

It would thus make a great deal of sense if the satellite and terrestrial broadcasters could agree on a joint approach to the protective enveloping of the net 16 Mb/s digital HDTV signal.

When the satellite signal is received and decoded at the cable head-end the resultant "satellite derived" television program signal should look exactly the same as an off-air terrestrial digital HDTV signal. It should have exactly the same appearance, use the same terrestrial protective envelope and should fit within the same 6 MHz bandwidth. In effect we are suggesting that the net satellite television signal (16 Mb/s) should be double-wrapped, first in the terrestrial envelope (the 18–20 Mb/s pack) and then in the thickly padded satellite envelope (the 22–30 Mb/s pack). This is shown in Figure 3-5.

For the purist this approach must appear bad news, for theory suggests that a single thick wrap (from 16 to 22 or 30 Mb/s)

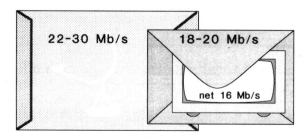

**Figure 3-5** Wrap compatible. The 16 Mb/s HDTV signal is first
enveloped in a terrestrial/cable wrapper before under-
going the final satellite wrapping process. This double
wrap allows an easy transposition from satellite to
cable transmission.

is much more effective than the two separate wraps proposed
above. However, we must take care that a neat theory does not
lead us too far astray from our primary objective, a digital HDTV
signal that is off-air friendly, cable friendly, satellite friendly and
VCR friendly. We must not repeat the HD-MAC mistake of being
able to get the signal from the satellite to the cable head-end but
no further.

We have thus established that, with care, a digital High
Definition television service can be designed to be compatible with
standard 6 MHz terrestrial and 24/36 MHz satellite broadcasting
techniques. This is shown in Figure 3-6. It can also be easily
routed via conventional radio frequency cable television wires and
telephone company fiber optic cable. However, before the digital
signal finally reaches the screen of our brand new high definition
television receiver there is another important entertainment
medium that warrants our due consideration—the domestic video
cassette recorder, the VCR.

**The HDTV videocassette recorder**

The digital videocassette recorder will eventually be less
expensive than its conventional analog cousin because it is a
much simpler concept and design. It is not required to record the
relatively fragile high quality analog signals, only a rugged standard speed 16 Mb/s digital signal. It seems probable that the need

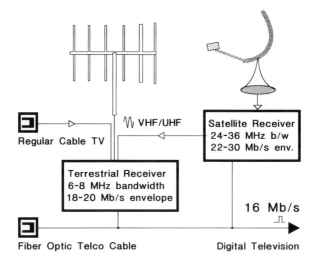

**Figure 3-6**  Digital compatibility. The same digital HDTV signal
may be broadcast from a satellite or terrestrial transmitter and routed via cable TV or telephone company
fiber cable.

for much of the watch-maker mechanical precision of current
conventional VCRs can be avoided and the cost can fall.

But this 16 Mb/s recorded signal is not just another "video,"
it is the introduction of real high definition television into our
home.

Broadcasters thus face the challenge of intense competition
from inexpensive, robust technology, high definition videocassettes which may contain programming of recent movies or sports
events and which can be rented or purchased from any store.
Unlike today's popular convention analog VHS video recording
format the digital VCR picture quality is no longer any worse than
an off-air or cable broadcast.

Furthermore the digital VCR tape can be copied without any
loss of quality, whether from an off-air source or from another
digital tape. Depending on our point of view this domestic recording facility is either a tremendous advantage or spells commercial
disaster.

How can this feature be tamed?

The designer of the domestic videocassette recorder has a number of interesting choices before him:

*Trick modes.* Will the domestic market accept a limited range of forward and reverse "picture search" speeds or is there an expressed consumer preference for a wider range of search speeds? The digital HDTV 16 Mb/s format may provide an easy implementation of a picture search facility at, say, 6 or 8 times life speed but other picture search speeds may be difficult to obtain. Does this matter?

*Compatibility.* Is it better to record the 16 Mb/s HDTV signal in its regular off-air 18–20 Mb/s form or is it better to invent a new form of enveloping which is more suited to the magnetic tape environment?

*Omissions.* Is it necessary to record the whole signal or can the digital decoder circuits tolerate short breaks in the data stream? At present many conventional inexpensive VCR players do not record all the invisible lines at the top and bottom of the picture (the frame synchronization pulses) yet still offer a perfectly adequate playback. It seems probable that the digital VCR might provide the same economy in design provided that we first arrange that some regularly recurring parts of the 16 Mb/s data stream can be safely discarded.

**Pay per play**

Up to now we have considered the digital video signal as composed of two parts. The first part is a sequence of some 5 to 10 frames per second which is provided in "absolute" form. In the first chapter we suggested that parts of this absolute frame sequence might be made particularly easy to decode so as to provide rugged clues for a picture search facility in a digital video recorder. The second part is the interpolated sequence of "difference" frames which may be first decoded and then interpolated at 50, 60 or 75 pictures per second.

In order to defeat the threat of illegal copying of digital videocassettes we might therefore elect to continue to broadcast the rugged "absolute" signals in a straightforward manner whereas the interpolated difference signals would first need to undergo some form of "conversion" or unscrambling process before they could be used. This is shown in Figure 3-7. The keys to this unscrambling could be deliberately contained in the part of the digital waveform that the domestic video recorder is unlikely to be able to record faithfully, just as the conventional inexpensive domestic video recorder is unable to faithfully record all the invisible lines in the TV picture. Such a technique might become more firmly established if we always ensure that any keyword information is always placed in the same "hard-to-record" spot in the digital HDTV signal.

It is now up to the broadcaster or, in more general terms, the program publisher to decide whether the program is to be distributed in plain or scrambled form. If the program is broadcast or otherwise distributed in a scrambled form then the videotape may still be recorded successfully on a domestic video recorder. However, it will not be possible to play it back correctly unless the previously deleted or missing keywords are re-inserted in the right place. In this way it is possible to levy a charge on the videocassette viewer for the use of the programming. This may be done on a time basis or, perhaps, each time the tape is played.

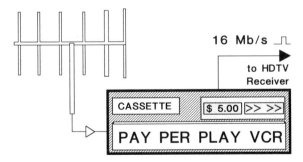

**Figure 3-7**  Pay per play. Digital recording is permitted but each playback requires a keyword or "token" to unscramble the picture.

## SEVEN SCENARIOS FOR DIGITAL TELEVISION

During the course of this book we have been beguiled by digital television as we have been led gently by the hand into new pastures. Thanks to the American research effort into HDTV the techniques for the digital transmission of television pictures are being transformed into electronics hardware and our expectations are running high. Everything seems possible; the high-speed Fast Fourier Transform computer chips are now available and the American taboo channels are no longer safe from the prying eyes of broadcast entrepreneurs.

The stumbling block for digital television is not the digital technology *per se* but rather its rate of penetration into the market. For us to envisage digital television as the broadcast panacea of the year 2,001 is one thing but to see how we might trace a profitable path from here to there is quite a different proposition. As Commissioner Duggan said in early 1993, "It should not be a death march for the broadcasters."

It is therefore worth trying to map out a scenario in which both satellite and terrestrial digital television transmission can be gently introduced to the marketplace, one profitable step at a time. The following pages discuss several different ways in which digital television transmission can be introduced to serve our existing NTSC or PAL television sets. It is only when this network is both profitable and established that we will enjoy a sufficient flow of funds which might permit the more widespread introduction of high definition television.

### Sharing the resources

At the end of the first chapter (page 89) we discovered that the 14 Mb/s video portion of the 16 Mb/s video/sound/teletext data stream could be split a number of ways. We agreed that we might squeeze two 700 pixel "Rec 601" quality pictures into the 14 Mb/s video data stream but that other possibilities also existed. If we choose to move up-market then we have seen how the 14/16 Mb/s data stream can, at a pinch, support a single 787½ line "high definition" television picture.

On the other hand we can just as easily settle for more TV

programs at lower quality, for our earlier calculations indicated that four or more NTSC quality pictures can be squeezed into the 14/16 Mb/s data stream.

In order to achieve four separate programs within the overall 16 Mb/s data stream it would appear that, in round figures, we need to allocate about 4 Mb/s apiece to each of the separate programs.

But is this enough?

We are aware that some TV scenes contain a lot of movement or fine detail and thereby occasionally need a higher than normal data rate whereas other scenes are far less onerous to encode.

It is at this point that the earlier analogy of the Chinese restaurant dining protocol can come to our rescue for we know that a flexible method of resource sharing, whether in the allocation of Chinese food or of digital data capacity, is far more effective than a fixed allocation method. Thus it is quite practical to squeeze or "multiplex" four or five NTSC quality television pictures into one 16 Mb/s data stream when each picture is permitted to take as much data capacity as it requires.

In the jargon of the data communications engineers this flexible sharing technique is termed "statistical multiplex" or "stat-mux" for short. The meaning of the "multiplex" term is straightforward whereas the "statistical" qualifier reflects our pious hope, our gamble even, that we will rarely encounter the embarrassment in which all the TV programs will want to broadcast "difficult" pictures at exactly the same time.

Before we start our Chinese banquet, however, it is important to know how many dishes have been ordered and how many people are coming to dinner.

### A terrestrial banquet

The preceding discussion has assumed that the overall resource is a channel that can support a digital transmission rate of about 16 Mb/s—the sort of channel that the FCC might soon allocate to every existing terrestrial broadcaster in the USA. If the broadcaster were to offer a number of separate NTSC programs within that digital channel then it is clearly his responsibility to sort out the sharing protocol. In other words any dynamic re-allocations that may occur must be contained within the overall limit of the 16 Mb/s digital channel.

### A satellite banquet

It seems probable that digital satellite broadcasts can support two quite separate 16 Mb/s digital channels within an overall transponder multiplex of 45 Mb/s—not one but two tables of food. In theory the satellite broadcaster can claim that it is more efficient for him to consider the two 16 Mb/s channels as a single 32 Mb/s resource. He may be tempted to dynamically re-allocate resources across the two 16 Mb/s channels which would allow him to squeeze in more NTSC programs, just like moving dishes between the tables.

### An ATM buffet banquet

A few pages back we learned that the telcos' ATM service could carry data streams that totaled about 110 Mb/s—the equivalent of seven 16 Mb/s digital channels or a total of about 35 NTSC programs. If these 35 programs emanate, five at a time, from seven terrestrial broadcasters there is no problem of dynamic reallocation—for these issues would have been dealt with by the broadcasters themselves.

However, if the ATM service intends to provide a wide ranging "video dialtone" choice of programming then problems can arise and pictures can get corrupted. The 35 dialed-up NTSC pictures will no longer emanate from seven broadcasters who have each addressed the dynamic re-allocation issues of their individual five-way programming. By analogy we have lost the discipline of allocating food between the five—some hungry, some less hungry—NTSC guests at each small table. We now run the risk of 35 very hungry NTSC guests descending on the central buffet all at once.

Unlike the broadcaster, who can briefly soften program 2 in order to accommodate some fast action in his program 3, it is too expensive for the telco to re-code each video. It is much simpler for the telco to allocate some "headroom" in the ATM data transmission in order to accommodate most of the expected variations in demand. Occasionally, however, too many ATM subscribers will dial up fast moving programs and the service will crash.

### The hamburger alternative

If we cannot agree on how to share the excellent food at the Chinese banquet then perhaps we should settle for something simpler. As we know, there is a ready market for good quality standard-size snacks at a standard price. The video CD will be such a product. CD technology has fixed the data rate at 1.5 Mb/s and, for the present, there is no more to be had. Just like hamburgers the fixed rate NTSC pictures are not brilliant but are more than adequate for most occasions.

A technical alliance between the video CD jukebox and an ATM video dialtone service could prove a winning combination. The well-behaved 1.5 Mb/s output of the pre-recorded video CD will cause the telco few headaches. Seventy video CD channels can be stacked up safely on each ATM line with no worries about "headroom."

## Scenario 1: Multi-program normal definition digital television

The left side of Figure 3-8 shows how identical digital television signals may be received from a number of sources: off-air terrestrial broadcasts, conventional or fiber optic cable television systems and satellite transmissions.

The ways in which we may achieve a good measure of cross-media compatibility were described more fully in the previous section entitled "Competing Media" so, for the sake of clarity, Figures 3–8 to 3–16 have hidden this technical complexity in the vertical rectangular box entitled "Convert." After this conversion to a standard 16 Mb/s format the digital signal passes to the video cassette recorder and the digital television decoder which are shown in the bottom right side of Figure 3-8.

The digital video cassette recorder is quite dumb.

It records the 16 Mb/s primary data stream with little regard to its content. Quite by default it thus records whatever mix of separate programs might be contained in the data stream.

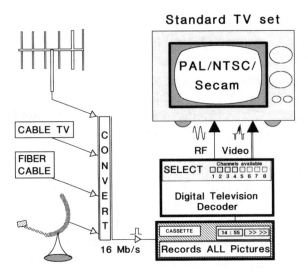

**Figure 3-8**   Normal definition. The digital television decoder is able
to select any one of a number of separate programs
which may be contained in the 16 Mb/s data stream.
The decoder output can be fed into the antenna socket
of a normal television receiver or to a baseband video
socket of a modern TV set.

The digital television decoder, on the other hand, is much
smarter.

It is able to recognize how many separate television programs are embedded within the overall 16 Mb/s primary data
stream. As we have already seen, it is technically possible to
embed five separate live NTSC television programs or eight
pre-recorded NTSC VHS-quality programs within the basic data
stream. In order to discover the exact number we can fit a set of
indicator lights on the front of the digital television decoder. The
viewer is able to select any of these programs one at a time.

> The selected program output of the digital television decoder
> is presented in the form of a standard NTSC signal so it can
> be seen on a regular NTSC television set.

The new "digital" program can be viewed by either tuning
the TV set to a spare channel, just as we would usually view the
output of the video recorder, or by routing the video signal to

the optional baseband video sockets that can often be found at the back of many modern television receivers.

Thus, for every 16 Mb/s data stream we can find on the dial, we may enjoy five or eight new TV programs.

But terrestrial broadcasting is not the only winner—conventional cable television systems can be re-configured to enjoy the same advantage.

Many "obsolete" 30 channel cable television systems can be brought back to life if they are re-configured to carry thirty 16 Mb/s primary data streams instead of thirty conventional NTSC transmissions. As each primary data stream is able to support four or five NTSC quality programs the aging cable system is now able to compete successfully once more by offering its subscribers the wider choice of 150 separate television programs.

In practice, however, the cable operator may prefer to convert just some of the less popular channels to digital form. He might re-configure only the final ten of the old 30 channel system in order to provide 50 separate nitch-iterest or premium business channels.

## Scenario 2: A helping hand—deaf sign language

During the course of the 1990 Broadcasting Bill through the UK parliament no one who was associated with this legislation could fail to notice the effective lobbying efforts of the Deaf Broadcasting Council. This comprised a truly delightful group of committed men and women who fought long and hard to alleviate the effects of their tremendous disability.

In the course of their campaign through the Commons Standing Committee they were finally able to persuade the government that the use of closed captioning—subtitling—in television programs was of such immense value to those with impaired hearing that the provision of such a service should become far more commonplace.

The 1990 Broadcasting Act requires every UK commercial television broadcaster to increase steadily the amount of teletext subtitling so that, by 1998, 50 percent or more of all television program hours are subtitled. Similar victories have been won for deaf people in the USA.

Although their first goal, closed captioning, was won amid great celebration, their second goal proved more elusive. Deaf signing—sign language—is tremendously useful for the deaf for it is to normal speech what closed captioning is to writing.

At critical parts of the Bill's progress through parliament the public benches in the Commons and the Lords were often filled to overflowing by a group of well-dressed but strangely silent men and women whose only contact with the proceedings was a sign language interpreter—for they were all totally deaf.

It was a delight to see how the normally stern parliamentary ushers were pleased to set aside their strict rules of "who sits where" in order to allow the signing interpreter to stand, facing them in silence, in the no-man's-land between her audience and the proceedings.

Although the government could thus see the importance of deaf signing at first hand, there was no easy technical solution to its widespread provision in television programs. Some UK television programs are already "signed" but for every viewer who welcomes this "overlay" in the corner of the television screen there are others who find this intrusion a distraction.

What is needed is the pictorial equivalent of closed captioning in which the viewer can turn the captions on and off at will. This is shown in Figure 3-9.

> But we already know that a pictorial form of digital teletext
> is no more than a description of digital television.

This technical similarity was not lost on the government advisers. In the House of Lords on July 24, 1990, during the passage of the Broadcasting Bill, the government Minister Lord Ullswater stated that:

> In the future digital television could be a very real possibility, though regrettably not soon enough for it to be a real consideration in this Bill. Such a development could revolutionize the use to which the broadcasting spectrum could be put and open up a whole plethora of additional opportunities for broadcasting and ancillary services.

I am told that with the development of digital television it will be possible, without using any additional space in the broadcasting frequency spectrum, to have one television picture superimposed on top of another. Think of the possibilities that will bring. It would be possible for those who wish to receive a sign language service simply to switch on the overlay picture, while those who did not wish to see it need not do so.

*Columns 1342–3 Hansard (Lords) July 24, 1990*

## Scenario 3: Wide aspect ratio screens with normal definition pictures

If we are to believe the television industry marketers then the imminent arrival of the wide aspect ratio television screen is to be the most significant new development since the arrival of color television. As the screen shape is quite different from what went before the rules for filling the screen may need to be re-drawn.

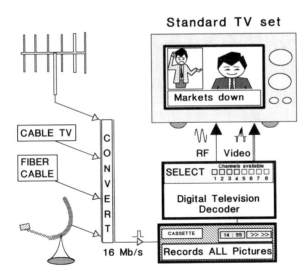

**Figure 3-9** Deaf signing. The shot of the sign language interpreter can be superimposed on the main picture at will, just as closed captioning (also shown) can be turned on and off.

There are two complementary areas of concern:

1. How best to fill a conventional 4 by 3 screen with the wider 16 by 9 transmitted picture.
2. How best to fill the wider 16 by 9 screen with a conventional 4 by 3 transmitted picture.

A considerable amount of effort has gone into researching the first option. As we have already seen, the Europeans have proposed wide-MAC and PALplus for 625 line receivers and the Americans and Japanese have proposed similar schemes to add side panels to their basic 4 by 3 525 line pictures. The simpler compatibility schemes propose some form of "letter-boxing" whereas different varieties of wide-MAC are able to provide full screen 4 by 3 pictures by simply discarding the wider aspect ratio side panel information.

Digital television can adopt exactly the same wing-dropping technique as wide-MAC.

If the digitally encoded picture is transmitted in a wide-screen 16 by 9 525 line format the decoded picture may be sent in its entirety to a suitable 525 line 16 by 9 display screen. On the other hand, if the picture must be displayed on a conventional 4 by 3 screen then the side panels that made up the original wider picture may be discarded.

The second area of concern, that of filling the 16 by 9 wide screen with a conventional 4 by 3 picture, also taxes the imagination to the full.

There are a number of options. The simplest approach is to sit the conventional picture slap bang in the middle of the wide screen but this results in two empty vertical bands that can be seen on each side of the picture. The overall effect is not unlike that of a letter box tipped on its side and is just what we set out to avoid on the conventional 4 by 3 screen.

If we are more ambitious then we can try to fill the empty space on the screen with some more interesting program material.

16 Mb/s digital television is ideal for our purpose because the primary data stream already contains other programs which are just waiting to be decoded and displayed alongside the main picture.

We need to turn back only as far as the last scenario in order to find a suitable candidate to fill the voids which occur when the conventional picture is shown on the wider screen.

We can easily arrange to display the "overlay" picture of the deaf signer either to the left or the right of the main picture. In this way we do not need to encroach on the main action. The result is shown in Figure 3-10.

A more upbeat example of what digital television technology might provide is illustrated in Figure 3-11. It shows how wide-screen stereoscopic television can be brought to the shopping mall or even our own living rooms.

The 16 Mb/s data stream carries two separate pictures:

1. A wide-screen stereo-left view
2. A wide-screen stereo-right view

The third picture, the "mono" screen may be easily derived from either of the two wide-screen stereo images by simply discarding the additional side panel information in the picture.

Since the birth of popular photography more than 100 years

**Figure 3-10**  Wide-screen signing. A wide aspect ratio screen can be used to display the "signing overlay" side by side with the main picture rather than on top of it.

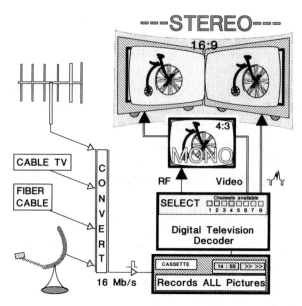

**Figure 3-11** Stereoscopic television. The data stream carries both left and right wide-screen pictures. The mono picture may be derived from either source.

ago various inventors have sought to demonstrate stereoscopic picture displays. Some techniques require the viewer to wear some form of special spectacles whereas other techniques use arrays of micro-lenses (lenticles) in front of the screen to produce the same stereoscopic effect.

It seems certain that any move away from today's cathode ray tube television displays to bigger and brighter liquid crystal displays will stimulate further interest in the manufacture of display screens that can produce stereoscopic images.

It is worth noting that although the 16 Mb/s data stream can normally support three independent wide-screen pictures this example has restricted the number to two. The two separate views that make up the stereo picture are very much alike and will therefore make the same demands for additional data capacity to cope with "difficult" pictures at exactly the same time. In this instance we are thus unable to take advantage of the statistical headroom that the transmission of independent pictures normally provides.


## Scenario 4: Bringing "Turner" down to earth

By mid-1993 less than 62 percent of the 93 million homes in the USA subscribed to cable. Despite the well publicized attraction of satellite television broadcasting it is difficult to envisage the day when the market penetration of satellite television service, whether delivered via cable or backyard dish, will reach 70 percent of US homes. Twenty eight million American homes will remain unserved.

> We take it for granted, however, that terrestrial broadcasting will fill the gap—after all, terrestrial transmissions reach over 99 per cent of the US population. In the UK the terrestrial penetration figure is even higher.

Satellite broadcasters must envy the ease by which the mass markets can be reached by the terrestrial broadcaster. For a start the transmission technology is more robust. We know from experience that VHF and UHF terrestrial television signals are relatively unaffected by the presence of trees next to our homes whereas the slightest obstruction in the line-of-sight can render the satellite television signal useless.

Digital television can provide a neat answer:

We know that the 16 Mb/s data stream can provide four or more NTSC quality television programs on a terrestrial taboo channel.

Why not let the FCC lease the access rights to a terrestrial taboo channel to Turner Broadcasting who currently provide a four program satellite television service across the USA. This is shown in Figure 3-12.

The leasing price of the taboo channel need not be exorbitant for it can reflect the fact that such a channel can efficiently provide a multiple program service at a transmitter power level

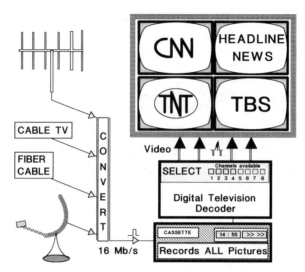

**Figure 3-12**    A terrestrial Turner. The Turner Broadcast company
could provide a four or five program terrestrial service through one 16 Mb/s data stream.

which is a fraction of that used by conventional analog high-power
terrestrial television transmitters.

Setting a leasing charge for the taboo channel brings us
naturally to the vexed issue of the pricing or auctioning of the
frequency spectrum. Although many high-power broadcasters
have expressed grave reservations about the introduction of such
schemes it would appear that a pricing mechanism based on both
bandwidth and power levels might help encourage a switch to
digital low-power terrestrial broadcasting. This has a beneficial
side effect.

The use of low-power digital transmitters makes the job of
the frequency planners much easier. As the power level of the TV
broadcaster is brought more into line with the transmitter power
used by the land mobile services we can afford to pull back on the
allocation of "guard bands." These are the wasted areas of un-
usable frequencies—the grass shoulders—that currently separate the TV bands from the adjacent cellular telephone services.
It is then up to the marketplace and the politicians to decide
whether these clawed-back frequencies should be allocated to
broadcasting or to land mobile service.

## Scenario 5: Exploiting the video wall

Every so often we get to see a really good audio-visual presentation. We have come to expect that the quality of the sound will be good but, just occasionally, the pictures can be really stunning as well. If we are allowed a quick peep behind the scenes we will discover that the clever "cross fades" and "dissolves" are achieved by a modern-day example of mechanical magic, the careful alignment of many slide projectors working in unison.

What a terrific sales aid—but what a difficult task to set it up properly!

The television equivalent of this super slide show is the video wall, a stacked matrix of television screens that can show a *collage* of still and moving pictures which is only limited by the artistic imagination of the video wall program producer.

The most stunning displays use more than one, usually four or six, simultaneous picture sources in order to provide novelty and variety.

Unfortunately the necessary control equipment is also quite complicated as it must include the four or six program sources that are usually taken from a set of video tape or video laser disc players. The idea remains attractive but the implementation is still too complex for more general use. Too complex, that is, until we use digital television to provide the source material.

Imagine a shopping mall or a gas station with an electronic billboard—a video wall—in full view of the public gaze. We know that to leave the day-to-day video wall programming in the charge of the shopping mall security guard or the filling station attendant is just asking for trouble, for even broadcasters can occasionally run the wrong tape. We should instead assign the day-to-day video wall programming to the experts, a central television advertising company who have contracted with their clients what promotions must be shown where and at what time. The shopping mall or the gas station will receive the multi-program feed via a digital satellite link from the central advertising "roll out" facility.

If this concept is to work it must be done well.

There is no longer any room in the marketplace for the fuzzy pictures that we see when standing in line at the bank. The pictures must be interesting, bright and flicker-free.

We might even re-introduce a touch of flicker, just to catch the public's attention and thus render the pictures simply irresistible.

If this is all too exotic what of the simplifications?

We know that the digital video recorder will record all the pictures in the 16 Mb/s primary data stream. Let's record a few hours of high quality video wall-to-wall promotional material on one cassette tape and then play it back time after time.

Is there a market for a domestic video wall?

Figure 3-13 illustrates how a sports program (in this case a penny-farthing bicycle performing "wheelies") may be presented on four screens in order to satisfy the sports fan who is determined to miss none of the action. If he should blink unexpectedly and miss a few seconds of the play he can always turn to his digital video recorder which has recorded every shot. Once recorded this

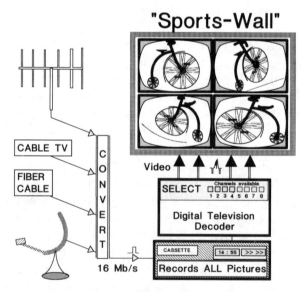

**Figure 3-13**   Sports wall. Four separate pictures of the same event can be shown on four separate display screens or reduced to a single screen. The video recorder captures all four scenes for later analysis by the viewer.

multi-shot playback is also available to any one with a standard television set.

## Scenario 6: Televising the UK parliament—MP-TV

The Committee corridor in the Palace of Westminster can be a busy place. Much of the Parliamentary Committee work escapes attention but some legislation can attract enormous numbers to the public galleries. The 1990 Broadcasting Bill which re-organized UK commercial broadcasting was of close interest to many and the 37 working sessions of the Bill's Standing Committee regularly drew large crowds.

Really large crowds?

Yes, on some days there were more than 60 members of the public lining up patiently for a place in the public gallery of Committee Room 10.

Each Tuesday and Thursday the Committee started work at 10:30 in the morning but it was pointless to arrive within less than 30 minutes of opening time. By 10 o'clock the line for the public gallery would often stretch across the upper vestibule and down the wide stone stairs.

The need for an early arrival in the line soon became apparent. The keen committee watchers arrived early—a 9 o'clock arrival might just guarantee one of the first six places in the line. Even then no one could be quite sure of a place.

"How," we all asked ourselves, "did the other guys get through the tight security so fast?" Had they acquired House of Commons passes or had they started from home even earlier than usual? How did they know in advance that a large contingent of line-swamping lobbyists was expected from commercial broadcast companies or the Deaf Broadcasting Council?

From 9 o'clock until 10:20 am the front of the line would consist of no more than a row of battered briefcases and old raincoats. The briefcase owners were otherwise engaged; they were conducting their business with friends and colleagues while sitting on the vestibule benches or quietly waiting for the public telephone to become free.

At 10:25 the waiting masses were marshaled into some sort

of order. The briefcases would re-acquire their owners and the counting would begin.

"The first twenty this way," would come the call, and the front of the line would shuffle forward towards the half-open door of the Public Entrance to Committee Room 10.

The first barrier had been passed.

Once inside the door another test of skill lay ahead.

The policeman, for there was always the same Scottish policeman, did his best to direct everyone to the two rows of seats at the back of the room in an orderly fashion.

How many seats would there be today and where should we sit for the best view?

We soon discovered that the number of seats was inversely dependent on the popularity of the particular clause of the Bill which had been scheduled for the day's discussion.

Administrative details in the Bill, such as the setting of provisions for pensioned-off broadcast regulators, rarely attracted any general attention whereas the more basic issue of defining the quality of television programs attracted a great deal of interest.

Although none of the broadcast lobbyists dared complain, important clauses also attracted the attention of the television cameras.

Really contentious clauses in the Bill would attract the devout attention of a three-camera crew. One of the television cameras would always be placed in the public gallery with the result that three seats would be forfeit. On many occasions the television picture mixing console was also sited alongside the camera in the public gallery with the consequent loss of another three seats. The regular waiting line members soon learned that, for popular clauses, there might be as few as twenty sets available in the public gallery.

On the other hand, there might be a few seats to spare during the less contentious parts of the Bill. Hence the need to be early in the line. The first twenty would always get into the Committee Room while most of the remainder would be turned away.

"But where do the others all go?" I asked a BBC lobbyist one morning as we both squeezed into the last two seats available.

"That's easy," she replied. "They return to the office and listen to the proceedings on the internal distribution system. That's what I do if I am turned away."

Until then I had not realized the full extent of the advantage that the broadcasters enjoy over the general public. They can listen in comfort to the proceedings from every committee room and nearly every microphone in Parliament just at the turn of a switch.

But where does it all go?

Only a little of it gets broadcast. Twice a week the mid-afternoon Prime Minister's Question Time is broadcast "live," together with 30 minutes of House of Commons business that follows it. Of the evening debates in both the Commons and the Lords there is not a trace.

Agreed a few snippets may make their way into the following morning's show of parliamentary highlights but the overall effect is lost. We are back once more to the topic at the beginning of this book where we discovered the need to first don our swordsman's mask before we could properly understand what is going on.

"Sound bites" are nowhere near enough to engage our attention, we need a continuum, a stability, in which we can more fully understand the legislative work being done on our behalf.

Many Members of Parliament share this view. When the Commons finally agreed to the experimental televising of its proceedings 64 MPs voted against the experiment because the proceedings could be not be shown in full—gavel to gavel.

Figure 3-14 illustrates how we might use digital television to broadcast pictures from both the House of Commons and the House of Lords together with simultaneous coverage of three committee meetings by adding these signals to the normal output of a digital News Channel.

The marginal cost of a full parliamentary television service for the general public would appear to be quite modest for the proceedings are being televised anyway. Figure 3-15 illustrates how a "videoplex" format will allow the interested viewer to follow a number of parallel parliamentary activities on a conventional NTSC television set. The larger center picture (shown as "Floor Leader") is the viewer's own choice made from a selection of one

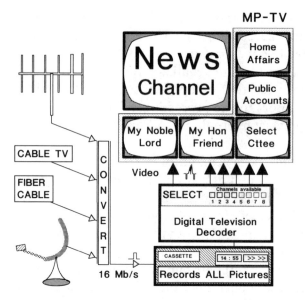

**Figure 3-14**   MP-TV supplements a News Channel. The relatively
static pictures from Parliament require lower data
rates than fast moving scenes. Five or more MP-TV
pictures can be added to the News Channel picture.

of the peripheral pictures. As always the digital video recorder
will capture every strand of the proceedings.

### Scenario 7: The penny farthing principle

In the earlier section of this chapter, entitled "The American
Class of 1991," Richard Wiley is quoted as saying "Four of the
proponents now present simulcast HDTV system concepts, em-
ploying an all-digital transmission format."

Good news indeed except for that qualifier, the jargon word
"simulcast." This word is a concatenation of two longer words,
**simul**taneous and broad**cast**ing and, for the American broadcasters, its meaning is straightforward. In the eyes of the existing
TV broadcasters the new technology, digital television, will allow
them to duplicate their existing program material on the newly
found taboo channels at higher picture definition and on wider
screens.

Non-broadcasters view these developments in a completely
different light.

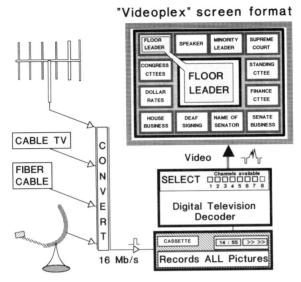

**Figure 3-15**   Full coverage MP-TV. Twelve or more parliamentary
pictures can be carried in the standard 16 Mb/s data
stream. The keen parliament watcher can follow all
of them on a standard TV set if a "videoplex" format
is used.

Radio interests argue that some of the new-found taboo
channels should be assigned to radio rather than television broadcasting. They question the need to place the proposed new T-DAB
(Terrestrial Digital Audio Broadcast) radio service in the microwave bands when a VHF television taboo channel or two would
serve their purpose very nicely.

Other prospective broadcasters question the assumption—
implicit in the word "simulcast"—that the new spectrum will be
assigned to the existing broadcasters. Surely, they argue, some of
the new-found spectrum should be assigned instead to ethnic
minorities whose voice is rarely heard on the air waves.

Fortunately terrestrial digital television allows us to recon-
cile both views.

As we have already seen, each taboo channel can support a
number of separate television programs. We can therefore assign
some of these program channels to the existing broadcasters and
the remainder to other interests.

Even if the existing broadcaster is allocated sufficient space within the 16 Mb/s data stream to provide a "simulcast" normal definition wide-screen version of the standard television program this still leaves 50 percent of the data capacity unused. This spare capacity could now be shared between the existing broadcaster and new entrants to the market.

In this way we can achieve two goals: an improved television picture from the existing broadcaster and a means of entry for the newcomer.

But we are faced with a problem.

This technique works well for digital television data streams that consist of a number of separate television programs but appears to fall down completely when the data stream is used to carry only a single high definition television picture. From what we have learned so far it appears that the HDTV signal will use up all the data capacity.

The new entrant broadcaster is thus faced with a very real threat to his business for his own program could be "bumped off" the data stream if and when the existing broadcaster decides to upgrade his transmission from normal definition wide-screen pictures to high definition wide-screen.

> It is important that this does not happen; the new entrant broadcaster must enjoy a good measure of security of tenure on the digital channel.

Hence the need for the Penny Farthing principle, in which the last 10 or 15 percent of the data capacity cannot be taken by the HDTV program.

This principle is illustrated in Figure 3-16 where the data stream is arranged to support a normal definition program as well as the high definition signal.

If this sharing principle is not firmly adopted we run the risk that the existing broadcasters will be tempted to "acquire" a neighboring taboo channel and use it in its entirety for high definition broadcasts. The Penny Farthing principle does not guarantee that minority programming will be successful but does at least give such programs a chance to move off first base.

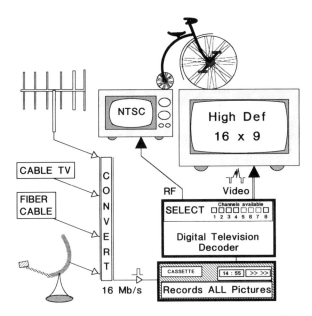

**Figure 3-16** Penny Farthing HDTV. The HDTV signal is not permitted to hog the whole of the data stream at the
expense of minority programming.

## OVERWHELMED BY CHOICE

Digital television has opened up a bewildering range of
opportunities. Some proponents at the quality end of the spectrum suggest that we should be setting our sights on 1,000 line
pro-scan systems whereas others believe that ultimate definition
4,000 line systems should be our eventual goal.

At the other end of the spectrum lies the Dirty Digital
DeathStar, a seething mass of 500 celestial TV channels. Take
your pick of the programming for it can range from a simultaneous showing of 500 different sub-VHS quality B movies to a
single Blockbuster movie that has been time delayed 500 times.

One hundred terrestrial channels, 500 cable channels and a
thousand satellite channels—all on a death march to a Chapter
11 filing.

Let's try again.

The 1992 Barcelona Olympic Games was a great television

occasion. Added to the extensive NTSC/PAL coverage of the Games there was a significant deployment of HDTV cameras. The European HDTV consortium, Vision 1250, broadcast a number of the events live and the consortium members who had acquired HD-MAC decoders were able to watch these experimental satellite transmissions on high definition monitors throughout Europe.

I watched them too. Not with an experimental HD-MAC decoder of course but on a regular PAL/D2-MAC $600 Ku band satellite receiver. In order to enjoy the pictures to the fullest I also purchased a 34 inch wide-screen television set which was fitted with a baseband video input.

The Barcelona D2-MAC pictures (and the more recent Lillehammer Winter Olympics pictures) which I watched in normal definition were excellent. In fact they were much better than the simultaneous PAL pictures being broadcast by the BBC. More confusingly, however, they were also much better than the usual D2-MAC satellite transmissions from France or Germany.

What was the difference?

The difference lay in the camera. The HDTV camera picks up four times as much detail as a normal definition camera. Even when the signal is later degraded to fit into a normal TV bandwidth the picture still retains an extra "something."

But there are many degrees of degradation.

The HDTV pictures which had been transmitted from the winter Olympic Games held earlier at Albertville had proved a disaster. The HD-MAC pictures had proved OK but the compatible D2-MAC pictures were appalling.

Why was this?

You will recall that wide-screen D2-MAC is a normal definition picture which is then "sharpened up" into its high definition HD-MAC form by the addition of a digital augmentation channel.

In the Albertville transmissions the raw signals from the HDTV cameras had been successfully transcoded into HD-MAC form. Or so it appeared at first. The digital augmentation had completely hidden the fact that the normal definition channel, the D2-MAC signal, was very jerky.

The good news from this experiment was that the digital augmentation was able to correct a poor picture.

The revised Barcelona HDTV transcoder therefore followed

a different protocol. The first step was to optimize the normal definition picture. Only then was a smaller amount of digital correction applied in order to produce the high definition version. By this means it was possible to produce good normal and high quality pictures simultaneously.

During the course of this book we have already encountered two similar examples.

In an ideal world we might prefer to enjoy 20 or 24 bit sound recording but for the present we must get by with the sound of 16 bit CDs. We have learned, however, that an intelligent "dithering" of the sound samples in the mastering process can make the reduction from 20 to 16 bits far less painful on the ear.

If a document enters a fax machine slightly askew then the received copy is sharp in parts but blurred in others. One way around such carelessness is to send the document in the fine resolution mode in order to reduce the amount of blurring. The price we pay is that the high definition fax transmission takes twice as long to send.

Could we perhaps invent a magic box that will accept a document on the skew, scan it in the fine high definition mode and then straighten it out prior to transmission at standard definition? With a little care our faxed paper documents might then look as sharp as a paperless transmission from a fax card in a personal computer.

The television market has already got such a magic box in the shape of the HDTV down-converter. It accepts an HDTV camera input and delivers a normal definition 525/625 line output that is much better than any normal definition camera could produce.

It was this HDTV down-conversion process that made me marvel at the quality of the D2-MAC pictures from Barcelona, even though my expensive 34–inch wide-screen set can barely resolve more than 500 horizontal pixels.

In the run up to the arrival of true HDTV in the home there is considerable potential for down-conversion. As the major networks gradually re-equip their television studios with HDTV cameras we can, through down-conversion, immediately enjoy the fruits of their investment on our regular NTSC TV sets at home.

Next, for a modest premium, we can purchase a normal

definition wide-screen television set that can offer us even better, wider pictures.

Just like the start of color TV 40 years ago we can at least promise ourselves that sometime in the future we will get around to true HDTV. In the meantime there must be a succession of affordable products which consumers can buy with confidence.


## STANDARDIZATION

We are still in the early days for digital television and there are still many questions to be answered.

- How many picture standards should we adopt and how many data rates are likely to be used?
- Should lower definition pictures be embedded in a higher quality picture or is it much simpler to offer two separate high and low definition transmissions alongside one another?
- Will multi-media be a real market or is there little substance in the computer industry's efforts to gate crash the TV business?
- Will viewers want to watch television when riding the freeway or is it sufficient to design systems that can work well with rabbit ears antennas?
- Will we want to see moving television pictures on our wrist?

If no one is able to answer Yes to these questions it is equally true that no one is prepared to say No.

Just like the Grand Alliance in the USA the MPEG (Motion Picture Experts Group) has found itself in such an invidious position when trying to set worldwide standards for digital television. It feels it has no choice but to take on every new idea that is mooted—for nobody yet knows who the winners and losers will be. Its meetings have been held in capital cities around the world and attract 150–200 interested parties, all with their own point of view.

Their task is unenviable but quite essential if digital television is not to be a very expensive flop.

I wish them every success.

# Bibliography

Anastassiou, D., and M. Vetterli, "All digital multi-resolution coding of HDTV," *NAB HDTV Conference*, Las Vegas, 1991.

Barnsley, M. F., and A. D. Sloan, "A Better Way to Compress Images," *Byte*, January 1988.

Chen W. H., C. H. Smith, and S. C. Fralick, "A Fast Computational Algorithm for Discrete Cosine Transform," *IEEE COM-25*, 1977, No. 9.

Citta, R., C. Eilers, R. Lee, and J. Rypkema, "The all digital spectrum compatible HDTV system," *NAB HDTV Conference*, Las Vegas, 1991.

Clarke, R. J., *Transform Coding of Images*, London: Academic Press, 1985.

Croll M. G., "Using the 8-bit CCIR Recommendation 601 Digital Interface," *IEE International Broadcasting Convention*, Brighton, 1988.

Daubney, C., "The technical co-existence of 4:3 and 16:9 pictures in the studio environment," *18th International TV Colloquium*, Montreux, 1993.

Davson, A., *Physiology of the Eye*, London: Churchill Livingstone, 1972.

DePriest, G. L., and G. M. Schmitt, "Advanced television, a terrestrial perspective," *IEE International Broadcasting Convention*, Brighton, 1988.

DiZenobio, D., "A double half-bandwidth OFDM system for digital video broadcasting," *IEEE Supercomm,* New Orleans, 1994.

Dobbie, A. K., "Why screened rooms in hospitals," *World Medical Electronics,* 1965, Vol. 3, No. 5.

Evans, B. T., "Ensuring a brighter future for local service DAB," *2nd Digital Audio Broadcasting International Symposium,* Toronto, 1994.

Evans, B. T., "The Future of Digital Television in the United Kingdom," *The Future of Broadcasting Vol. II—Minutes of Evidence and Appendices,* Home Affairs Select Committee, London, HMSO, 1988.

Farrel, J. E., "Objective methods for evaluating screen flicker," *Selected papers presented at the conference on work with display units,* North Holland, Knave & Wideback, 1986.

Forrest, J. R., and G. J. Tonge, "Television in the next decade, standards convergence or confusion," *IEE International Broadcasting Convention,* Brighton, 1990.

Gabor, D., "Theory of Communication," *J.IEE,* 1946, Vol. 93, Part III.

Gabor, D., "New possibilities in speech transmission," *J.IEE,* 1947, Vol. 94, Part III, No. 32.

Gross, L. S., (ed.)., "The international world of electronic media," New York: McGraw Hill, 1994.

Kelly, D. H., "Theory of flicker and transient responses 1. Uniform fields," *J. Optical Society of America,* 1974, Vol. 61, No. 4.

Lothian, J., B. Beech, and C. Baudoin, "MAC on Cable—the route to HDTV," 16th International TV Symposium, Montreux, 1989.

Lucas, K., "B-MAC and HDTV—does it fit?," *HDTV Colloquium,* Ottawa, 1987.

Mandelbrot, B. B., *The Fractal Geometry of Nature,* New York: Freeman, 1982.

Oppenheim, A. V., and A. S. Willsky, *Signals and Systems,* Englewood Cliffs, New Jersey: Prentice Hall, 1983.

Powers, K. H., "High definition production standards—interlace or progressive?," *SMPTE Annual TV Conference,* 1985.

Powers, K. H., "Framing the camera image for aspect-ratio conversions," *NAB HDTV Conference,* Las Vegas, 1994.

Quinton, K. C., "The carriage of MAC/Packet signals on cable systems using VSB frequency modulation," *16th International TV Colloquium,* Montreux, 1989.

Riemann, U., "PALPlus specification and principles," *IEE Colloquium "PALPlus,"* London, 1994.

Rypkema, J. N., "Spectrum and interference issues in ATV," *IEEE Trans. Consumer Electronics,* 1989, Vol. 35, No. 3.

Schroeder, M. R., "Models of Hearing," *Proc. IEEE, 1975,* Vol. 63, No. 9.

Schroeder, M. R., "Optimizing Digital Speech Coders by Exploiting Masking Properties of the Human Ear," *J. Acoustic Society of America,* Dec. 1979, Vol. 66.

Snell, R., "Techniques for HDTV Up and Down Conversion," *18th International TV Colloquium,* Montreux, 1993.

Swinson, P. R., "Film for HDTV," *IEE Colloquium "Film for television, alive or dead?,"* London, 1991.

Tonge, G. J., "Television Motion Portrayal," *IBA Experimental & Development Report, 134,* London, 1985.

Wells, N. D., "Bit rate reduction for digital TV," *IEE Colloquium Prospects for Digital Television Broadcasting,* London, 1991.

Windrum, M., "The future of satellite television—the digital option," *18th International TV Colloquium,* Montreux, 1993.

## Reports

"Digital sound programme transmission impairments and methods of protection against them," Geneva, *CCIR Report* 648-3.

"Encoding Parameters of Digital Television for Studios," Geneva, *CCIR Recommendation* 601-1.

"Ergonomics of Design and Use of Visual Display Terminals (VDTs) in Offices," London, *British Standards Institution,* 1990, BS 7179, 6 parts.

*United Kingdom Broadcasting Act,* London, HMSO, 1990.

"Visual Display Units," *House of Lords Select Committee on the European Communities,* London, HMSO, 1988.

# Index

**239**

Brian Evans obtained his PhD in medical electronics at St. Bartholomew's Hospital, London, after a first degree in electrical engineering.

After some years in telecommunications at British Petroleum, he decided that digital television as the technology of the future.

This book describes his researches and conclusions.